Crystal Growth

Edited by Vadim Glebovsky

Published in London, United Kingdom

IntechOpen

Supporting open minds since 2005

Crystal Growth
http://dx.doi.org/10.5772/intechopen.73360
Edited by Vadim Glebovsky

Contributors
Daniel Rasic, Jagdish Narayan, Jose Murillo, Jose Ocón, Guillermo Herrera, Jose Murillo-Ochoa, Gabriela Ocón, Vadim Glebovsky, Vladimir Kurlov, Sergei Rossolenko, Irina Shikunova, Dmitry Stryukov, Kirill Zaitsev, Irina Dolganova, Gleb Katyba, Kamila Ferreira Chaves, Thaís Jordânia Silva, Ana Paula Badan Ribeiro, Maria Aliciane Fontenele Domingues, Daniel Barrera-Arellano, Evgenii Nikolaevich Mokhov

First published in London, United Kingdom, 2019 by IntechOpen
IntechOpen is the global imprint of INTECHOPEN LIMITED, registered in England and Wales, registration number: 11086078, 7th floor, 10 Lower Thames Street, London, EC3R 6AF, United Kingdom
Printed in Croatia

British Library Cataloguing-in-Publication Data
A catalogue record for this book is available from the British Library

Additional hard and PDF copies can be obtained from orders@intechopen.com

Crystal Growth
Edited by Vadim Glebovsky
p. cm.
Print ISBN 978-1-83962-674-6
Online ISBN 978-1-83962-675-3
eBook (PDF) ISBN 978-1-83962-676-0

For EU product safety concerns:
IN TECH d.o.o., Prolaz Marije Krucifikse Kozulić 3, 51000 Rijeka, Croatia, info@intechopen.com or visit our website at intechopen.com.

Meet the editor

Prof. Dr. Vadim Glebovsky was born in 1936 in Ekaterinburg/Urals, Russia. After graduate and postgraduate study he obtained his Ing. Dipl. in Physical Chemistry and Practical Metallurgy from the Federal Ural University in 1961. He received his PhD (1963) from the Baikov Institute of Metallurgy RAS on scientific studies of purifying liquid metals by oxide melts using HF levitation. He was then invited to the Institute of Solid State Physics RAS (1969) as Chief of the Material Science/Physical Metallurgy Department, working mainly with growing/studying single crystals of high-purity refractory metals. He was habilitated (1985) with his doctoral degree for his research into the growing processes and substructure features of single crystals of W, Mo, Nb, etc. Since 1988 he has been Professor of the Institute of Solid State Physics, Chernogolovka, Russia.

As a multivisiting professor at the Technical University, Eindhoven, Netherlands, from 1992 to 2002, together with Prof. H.H. Brongersma and his students, Prof. Glebovsky studied the bulk and surface processes of single crystals of W and Mo by low-energy ion scattering. In the period 1994–1995 he visited the Lawrence Livermore National Laboratories (Livermore, California, USA), where, together with Dr. M. Campbell, they studied bicrystals of high-purity Mo. At the Max-Planck-Institutes für Metalforschung (Stuttgart, Germany), together with Prof. H. Fischmeister and Dr. P. Gumbsch, Prof. Glebovsky studied the cleavage fracture and brittle-to-ductile transition in W single crystals (1995), and later on, also at this institute, together with Dr. D. Brunner, studied the plastic properties of high-purity W single crystals (1998).

Nowadays, Prof. Glebovsky's scientific interests extend from certain aspects of the application of high-purity refractory metals and compounds as effective diffusion barrier layers in very-large-scale intergation to a new generation of high-sensitivity detectors made of low-radioactivity Ti for registration of "dark matter" particles.

Prof. Vadim Glebovsky has published more than 320 scientific articles and patents, one book in Russian on levitation melting, five book chapters, and has edited three books on various aspects of functional materials.

Contents

Preface

This book is intended to provide information on advances of new knowledge in the different technologies of growing single-crystalline materials to physicists, researchers, and engineers working in the field of crystal growing or applications.

Chapter 1 is an introductory chapter and is devoted to growing and testing W single crystals, which are oriented for single slip and have been tested in dynamical tensile tests at temperatures between 26 and 800 K. Critical shear stress τ_c at 800 K was not high ($\tau \approx 13$ MPa), which was consistent with both the high purity and structural quality of tested W specimens. The value of τ down to low T increased very quickly and parabolically. Qualitatively, these results agreed well with type A and type B tests of high-purity Mo single crystals. The measurements also confirmed the existence of three regimes for the dependence of flow stress on temperature for W single crystals.

In Chapter 2 the doping of SiC crystals during sublimation growth and diffusion is presented. The preparation of SiC crystals doped with various impurities introduced during the process of sublimation growth and diffusion is described. Crystals of n- and p-type conductivity with maximum content of electrically active impurities (of the order of 10^{21} cm^{-3}) are obtained. The solubility values of more than 15 impurities are determined. Special tantalum containers with several temperature zones, allowing the introduction of any impurity into SiC, are developed. The dependence of impurities concentration on temperature, growth rate, and seed orientation are found. Diffusion of impurities of boron, aluminum, gallium, beryllium, lithium, nitrogen, and phosphorus in silicon carbide polytypes is studied. Diffusion coefficients of these impurities in a wide temperature range are determined. Fast-diffusing states are atoms located in interstices, as well as centers, including the impurity atom and point defect. The extremely low diffusion mobility of lattice point atoms in the SiC lattice is noted.

Chapter 3 is devoted to the analysis of the behavior of the profile curves of the melt menisci for sapphire crystal growth by the edge-defined film-fed growth (EFG) technique. The menisci of the shaped crystals with capillary channels, fibers, and tubes (including cases of outer and inner circular menisci) are considered. Also, we investigate the profile curves of menisci in the cases of both positive and negative angles between profile curve and the working edge of the die. The cases of outer and inner circular menisci of the tubular crystals and menisci at capillaries and fibers are considered.

Chapter 4 describes growth of single-crystalline $LiNbO_3$ particles by the aerosol-assisted chemical vapor deposition method. Adjusting nucleation conditions, the effective shape and size control in the preparation of single-crystal lithium niobate nanoparticles by the aerosol-assisted chemical vapor deposition method is demonstrated. The effect of the most relevant parameters leading to nanocrystals taking a specific shape or size once they are synthesized is analyzed. This allows us to demonstrate that it is possible to control the size and morphology of particles prepared by adjusting the nucleation conditions. The synthesized nanocrystals show different morphologies, including quasi-cubic, tetrahedral, polyhedral, and hexagonal shapes, with characteristic sizes ranging from a few tens to a few hundred nanometers. However, rod-like structures with characteristic lengths ranging from 3 to 5 μm are also obtained. Electron microscopy techniques reveal the single-crystal nature of the synthesized particles.

Chapter 5 presents epitaxial thin film heterostructures, which are critical for integrating multifunctionality on a chip and creating smart structures for next-generation solid-state devices. Here, we discuss the traditional lattice matching epitaxy for small lattice misfit and domain matching epitaxy, which handles epitaxial growth across the misfit scale, where lattice misfit strain is predominant and can be relaxed completely, meaning that only thermal and defect strains remain upon cooling. In low misfit systems, all three sources contribute to the residual strain upon cooling, a result of incomplete lattice relaxation. In the second part of the chapter, we discuss the two critical contributors to the stress of epitaxial film: the thermal coefficient of expansion mismatch and the lattice plane misfit. In the last part of the chapter, the authors focus on unique cases where room temperature epitaxial growth is possible in nitride and oxide thin films.

Chapter 6 discusses the crystallization behavior of fats and oils, which is essential to ensure certain desirable characteristics in a specific industrial application. In recent years, some advances in the structuring of lipid phases have enabled direct influence on food properties. The structuring mechanisms of lipid bases can be classified as either conventional or unconventional. Conventional crystallization mechanisms consist of nucleation, growth, and maturation of the crystals, thus resulting in a crystalline lattice. Co-crystallization or seeding agents and emerging technologies such as ultrasound can be used to aid in crystallization and improve the physical properties of fats and oils. Unconventional mechanisms bring organogel technology as a trend, which consists in the use of self-assembly agents to entrap the liquid oil, resulting in a structured gel network. In this chapter, the formation process of crystalline networks and gel networks is presented in stages, highlighting the main differences related to the mechanisms of formation and stabilization of both types of network.

Vadim Glebovsky
Professor,
Institute of Solid State Physics,
Russian Academy of Sciences,
Russia

Section 1

Introduction

Chapter 1

Introductory Chapter: Growing W Single Crystals by EBFZM for Studying Mechanical Behavior

Vadim Glebovsky

1. Introduction

Tungsten (W) is one of the most perspective metals for different applications of its physical and chemical properties [1]. An incredible complex of diversified natural properties, such as mechanical properties, wear, and radiation resistance, stimulates a wide use of high-purity W single crystals in many modern applications, sometimes like quite unexpected ones, that is, W single crystals as high-resolution STM tips or elements of sputter composite magnetron targets for very-large-scale integration (VLSI) metallization [2, 3].

Electron-beam floating zone melting technique (EBFZM) is a unique technique for the crucibleless melting of such refractory metals as W, when it is contraindicated to have any contact with any refractory materials. This method is practically indispensable for the melting, refining, and growing of tungsten single crystals. As a result of numerous studies, it was established that the structure of tungsten crystals under the influence of large temperature gradients can differ from ideal. The author and his colleagues dealt with this structural problem for a long time. Single crystals of BCC refractory metals grown from the melt contain a lot of dislocations, so their density in regular samples can be up to 10^5 to 10^7 cm^{-2}. In many studies, it is shown that most of these dislocations aggregate into walls and grids, thus forming a characteristic dislocation substructure [4–6]. This chapter presents the results of the complex studies of the growth and mechanical properties of W single crystal by EBFZM depending on the growth rate, seed perfection, and axial temperature gradients. It seems that single crystals of W are the optimal objects for studying both growth processes and plastic deformation processes. I am confident that the studies presented in this chapter will contribute to further progress in this area. For several years, studies have been conducted in which attempts have been made to find out what prevents the growth of more or less perfect single crystals of W, on the one hand, and, on the other hand, attempts have been made to understand the patterns of plastic deformation of W over a wide temperature range.

Because single crystals of high-purity metals like W, Mo, Ta, and Nb, grown from the melt have the dislocation structure, characterized by blocks and boundaries thus, in our opinion, it is more correct to use such words as substructure, sub-blocks, or sub-boundaries. Several mechanisms for the appearance of dislocations and, accordingly, a characteristic substructure in the process of growing crystals are experimentally investigated before studying the mechanical behavior of W single crystals: under the action of the thermal gradients and mechanical stresses developed at the crystal growth, other factors have no such pronounced influence.

An investigation of the impact of a number of technological parameters of EBFZM on the substructure of grown single crystals has been carried out on the newly created growth equipment using fundamentally new electron guns allowing

IntechOpen

growing crystals of the optimal length and diameter in fully reproducible temperature conditions. The new electron gun is designed to form a stable circular electron beam. The main advantages of the gun are practically complete absence of a condensate of the melted metal on the circular cathode, fine focus of the circular electron beam on the crystal (anode) due to the focusing electrode system, the ability to grow crystals of different diameters through the use of a set of replaceable focusing electrodes, the stability and reproducibility of the temperature profile on the growing crystal, and an absence of warped elements in the design of the EB gun in the process of long-term work due to its fabrication of a water-cooled copper.

Nevertheless, the seed, growth rate, interstitials, and temperature profile on an interface have a significant effect on the final structural quality of grown crystals. In our studies, the chemical composition and temperature profile are kept constant due to the use of the original EB gun and pre-refined blanks, so it is always possible to distinguish the effect of the growing rate in a pure form. The role of the growth rate is of high importance in both the crystallization stage and cooling to room temperature (RT), starting beyond the interface between the liquid metal and the crystal. In bulk BCC metals, grown from the melt, deformation at RT is a rule controlled by dislocations with higher lattice resistance. Nevertheless, I am absolutely convinced that such a metal as W, which possesses such a unique complex of excellent properties, should be investigated, despite the enormous difficulties that must be overcome.

2. Growth of W single crystals and study of their plasticity

2.1 Introduction

The crucibleless melting methods are especially important in the production and investigation of single-crystalline W of high purity, since this metal has a high melting point and is chemically active in the liquid state. Along with cylindrical W single crystals, much attention is paid to W bicrystals, which are an extremely interesting object in carrying out various kinds of materials science studies, since it allows one to study grain boundaries with given crystallographic parameters. Although the structure of W single crystals has previously received quite a lot of attention, many questions arise as to the possibilities of improving the structural quality of W crystals, which requires a detailed analysis of factors that have a decisive influence on the formation of the structure. This explains the great interest to studies of ways to improve the structural quality of W single crystals depending on the growth parameters, namely, the rate of crystallization, electron beam intensity fluctuations, and the crystallographic perfection of seeds.

In conformity with the experimental results of dynamic tensile measurements, a form of the hardening curves depends on an orientation of a tensile axis of the specimens. For the orientation of the tensile axis near {1 1 0}, the stress-strain curves give an information on following small work-hardening, whereas for other tensile axes, a significant work-hardening after the elastic limit is discovered. As to glide systems, the {1 1 0} [1 1 1] glide system at low-temperature regime is operative, while at RT and above {1 1 2} slip planes also appear. In this chapter, dynamic tensile measurements on W monocrystalline specimens at T from 25 to 800 K are presented. The crystal axis orientation of tensile specimens has been chosen at the middle of the stereographic triangle because such a choice allows one to obtain the greatest stress in a {1 1 0} [1 1 1] slip system. As known, BCC metals at plastic deformation go through a ductile-to-brittle transition. The flow stress also strongly depends on T. The brittle behavior is noted in the cleavage form predominantly on the planes {1 1 0} and {1 0 0}.

2.2 Growth of W single crystals

Zone purifying of the starting material and growth of single crystals is carried out in a vacuum of 10^{-5} to 10^{-7} Torr at a power up to 25 kW, voltage up to 20 kV and cathode heating current up to 40 A. In accordance with the design capabilities of the setup, the diameter of W single crystals can be correctly varied from 4 to 30 mm. For effective melting and growing of single crystals, a fundamentally original electron gun has been developed (**Figure 1**). In fact, the electron gun is an electrostatic lens and provides rotation and focus of the annular electron beam. The gun is very reliable—the duration of continuous operation covers the most stringent technological requirements associated with refining an initial metal and growing single crystals. With the help of the original electron gun, it is possible to carry out long vacuum purifying and growing single crystals of W of any geometry, which is practically impossible when using the known guns.

During the EB zone vacuum melting, a deep purification of the liquid W occurs due to evaporation of volatile metallic and nonmetallic impurities. Thanks to this, it is possible to obtain W crystals with a very low content of impurities, often beyond the limits of detection by modern analytical methods (mass spectrometry with inductively coupled plasma, fast neutron activation, etc.). The stage of diffusion transparency of a liquid metal, when impurities diffuse extremely fast from the volume to the surface of the melt, is realized most effectively in the process of EBFZM in vacuum due to the phase transition of a metal from a solid state to a liquid one.

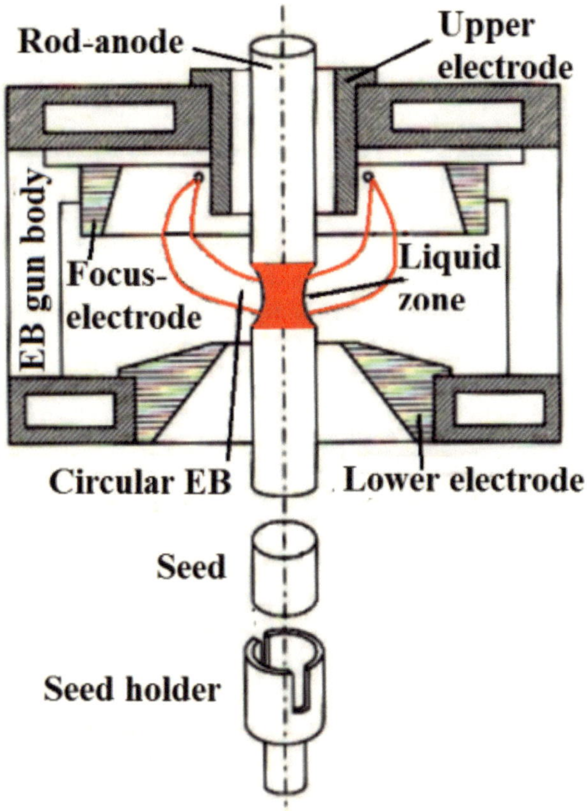

Figure 1.
A cathode unit (EB gun) for the EBFZM growth of W single crystals.

For this study, W single crystals having growth axes [1 1 0], [1 0 0], and [1 1 1] are of a diameter of 10 mm. The crystallographic orientations of both the seeds and growth axes are checked by Laue X-ray diffraction. Structural studies are done with both electron and optical microscopies. It is shown that all specimens are single crystals of the high quality, and their dislocation density is of $5 \times 10^5\ cm^{-2}$. The substructure has been also studied by the angular scanning X-ray topography which determined the position, size, and misorientation angles of subgrains. This technique consists in mapping an intensity of the diffracted beam over the crystal cross section in a fixed Bragg geometry. Samples after erosion cutting and mechanical grinding are subjected to electrolytic polishing and etching. The residual resistivity ratio demonstrates an integral purity of specimens and is measured by a four-contact technique. All W specimens used in our study have $R_{300K}/R_{4.2K}$ = 70,000. A typical chemical composition of high-purity refractory metals studied is given in Refs. [3–5]. Since the fulfillment of the tasks posed to a large extent depends on the structural quality of the crystals, the presence of the growth equipment for growing crystals with fully reproducible growth parameters is of great importance. It is already noted earlier that crystals of refractory metals have a characteristic dislocation structure with a subgrain size, which can be conditionally divided into three orders (**Table 1**). The structural quality of grown single crystals depends on the growth rate, thermal stresses, interstitials, inheritance of the seed structure, and supersaturation of the lattice with vacancies. The growth rate or travel of the liquid zone is one of the basic parameters. At a high temperature, dislocations, regardless of their nature, are very mobile, as a result of which the dislocation structure is polygonized. Along with the increased mobility of dislocations, a considerable contribution to the formation of a polygonized structure is also contributed by mechanical stresses. The growth rate of the crystals is from 0.2 to 50 mm min^{-1}. Usually, three passes of the liquid zone are necessary to get a crystal of both the high chemical purity and structural quality: N1 pass at 6 mm min^{-1}, N2 pass at 2 mm min^{-1}, and N3 pass at a given rate from the above interval. N1 and N2 passes are refining ones, and N3 pass is intended for growing a single crystal on a seed.

The structure of W crystals is somewhat different depending on the rate of the crystal growth. Crystals grown at a low rate (0.5 mm min^{-1}) are characterized by a developed substructure, with an average subgrain belonging to the second order (**Table 1**). The average subgrain size reaches 100 μm; the dislocation density calculated from the etch pits is of $3 \times 10^5\ cm^{-2}$. At a high growth rate (6 mm min^{-1}) the substructure contains separate etch pits without polygonal boundaries. The dislocation density is $\sim 1 \times 10^6\ cm^{-2}$. Another structural feature of W crystals grown at all rates is an inhomogeneity of the microstructure in the radial direction. The central part of a crystal is free of the subboundaries, and intensive polygonization can be observed at the periphery of the specimen. The block structure is most developed near the surface, and the boundaries of sub-blocks of the first order in the center of the crystal are absent; a misorientation angle of sub-blocks of the second order decreases

Order of substructure	Mean size of subgrains	Misorientation between subgrains
First order	1 mm < d < 8 mm	$30' < \Theta < 4^0$
Second order	50 μm < d < 1 mm	$30'' < \Theta < 30'$
Third order	0 < d < 50 μm	$0 < \Theta < 30''$

Table 1.
The substructure parameters of W single crystals.

substantially and the sub-blocks themselves are more equiaxed. These data indicate that the growth rate completely determines the nature of the substructure of the growing crystal.

The characteristic microstructure of W[0 0 1] crystals has been revealed on the (0 1 0) plane. This plane and the growth axis are parallel. The subgrain boundaries propagate over long length along the growth axis. They are dislocation grids left by dislocations having Burgers vectors $a/2[1\ 1\ 1]$ and $a[1\ 0\ 0]$, characteristic for the BCC lattice. The average subgrain size can reach 400 μm. The substructure of the W(1 1 0) single crystal is determined by the method of angular scanning X-ray topography. The subgrain is usually shown in various shades of gray; the light intervals correspond to the small-angle boundaries between them. Significant changes in the substructure are observed at the extra-high rates (>10 mm min^{-1}), when sub-boundaries appear in crystals with the misorientation angles up to 3–5°. It is experimentally established that the dislocation density increases with the increasing growth rate, however, not more than an order of magnitude. Thermal gradients are measured with a help of micro-optic temperature measuring as well as estimated with a digital solution of the heat equation for a stationary crystal growth stage. These estimates show that the axial gradients of T under the crystallization front in the solid phase can reach 1500 K cm^{-1}. This leads to significant thermal stresses followed by their relaxation through plastic flow and multiplication of dislocations. In other words, the temperature gradients cause high values of the dislocation density, which leads to the appearance of a characteristic dislocation substructure.

The appearance of dislocations having Burgers vectors a/2[1 1 1] and a[1 0 0] is very probable (a is a period of the BCC). It is precisely from these dislocations arising in the growing process of the crystal, due to high thermal stresses, that small-angle boundaries form in the W single crystals a characteristic dislocation substructure. At low dislocation densities of 10^5 to 10^6 cm^{-2}, small-angle screw or tilt boundaries are formed; however, they consist of several dislocation systems. The fine dislocation structure has been studied on the W single crystals subjected to a high-temperature creep. The small-angle boundaries have misorientation angles of 2–4°, and the dislocations are resulted of plastic deformation during cooling. In BCC lattices, symmetric tilt boundaries which consist of parallel edge dislocations are most likely in {1 1 1} and {1 0 0} planes, whereas asymmetric tilt boundaries are most likely in {h k o}-type planes and pure screw boundaries in {1 1 0}. Thus, if a dislocation system that creates small-angle boundaries in a plane layer is allowed to move by sliding or creeping, then the system of small-angle boundaries will tend to a finite number of planar grids. Based on this, the triple junctions of small-angle boundaries can differ from 120°. This is confirmed by our experiments on crystals which are grown at >1 mm min^{-1} (i.e., actually annealed). In this experiments, joints of different configurations are observed on the {1 0 0} planes. The use of recrystallized seeds of the highest structural quality, when the dislocation density is of 5×10^4 cm^{-2}, shows that boundaries germinate in a crystal. Even in the case, when the structural quality of seeds is very high, dislocations appear anew during further growing and can achieve values of 10^6 to 10^7 cm^{-2}. Due to polygonization the characteristic dislocation structure is formed, completely analogous to that in the regular crystals, with the misorientation angles of subgrains increasing step-by-step. The substructure of seeds is inherited by the crystals grown by EB zone melting. As a rule, boundaries presented in regular seeds also grow into a crystal. If the plane of such boundaries and the axis of a growing crystal are parallel to each other, then stable small-angle sub-boundaries persist in the crystal and sprout over long lengths. Thus, in the crystals with the growth axis [1 0 0], there exist small-angle boundaries elongated along the growth axis and having misorientations up to 1–2°.

A fairly high ratio of the resistances of the investigated single crystals of W at the level of ~10^4 to 10^6 indicates a high purity of the metal, which leads to an unambiguous conclusion about the insignificant role of impurities in the development of the crystal dislocation structure.

A comparison of two groups of W single-crystalline specimens has been done: as grown W crystals (AG) and strain-annealed specimens (SA). From the AG specimens, both groups of initial specimens are machined. An accuracy of the orientation of the final specimens, which depends on the structure quality, has been achieved: in the AG specimens, it is 1°, and in the SA specimens, it is better than 0.1°. To get a distinction of the structure quality of the SA and AG specimens, the characterization has been accomplished for both kinds of specimens. The AG substructures, influenced by thermal stresses, are detected after an electrolytic etching which identifies small-angle sub-boundaries separating the different subgrains. The substructure of the AG specimens W(1 0 0) contains low-angle sub-boundaries which appear as lines. An average size of subgrains is 500 μm. The subgrains are little stretched out in the parallel to the growth axes. The etching procedure reveals point defects and dislocations. It is found that the dislocation density is of 10^5 to 10^7 cm^{-2}. As for the SA specimens, it is not possible to use the etching technique because small-angle boundaries are absent. Thus, for a direct comparison of the AG and SA specimens, X-ray diffraction methods are used [7]. In **Figure 2a** and **b**, rocking curves are given for the AG specimen W(1 1 0) and the SA specimen W(1 1 0). It should be noted that in **Figure 2** the AG and SA curves are with different scales: **Figure 2a** is in angular degrees, while **Figure 2b** is in angular seconds. This method is revealed in the AG specimen several subgrains of different orientations as individual peaks which indicated by arrows in **Figure 2a**. The angles between the subgrains are calculated using data for the rocking curves, and they are of about 1°. The structural quality of the SA specimen in comparison with the AG specimen is obvious when one compares **Figure 2a** and **b**. The rocking curve for the SA specimen does not reveal any substructure. The rocking curve across the sub-boundary is presented in **Figure 2c**. Based on the fitting of this curve and two individual curves with the width taken from the SA specimen in **Figure 2b**, it is established that the misorientation angle between two subgrains is approximately 70 arc seconds.

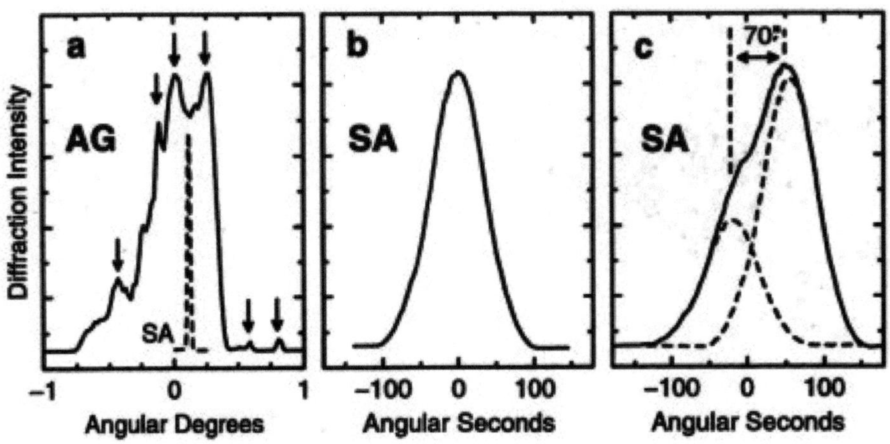

Figure 2.
Rocking curves of W(1 1 0) specimens: (a) AS specimen, (b) SA specimen, (c) SA specimen with small-angle sub-boundary; (a) also shows the rocking curve of the SA specimen taken from (b) for comparison (dashed curve). Arrows in (a) indicate different subgrains. The rocking curve (c) is measured across small-angle sub-boundary.

2.3 Surface cleaning and characterization of crystals

Surface cleaning and characterization of the grown W crystals are done in the UHV setup MiniMobis (1×10^{-10} mbar) [7]. To analyze an outermost atomic layer of specimens, we have used low-energy ion scattering (LEIS). These studies are performed with 5 keV $^4He^+$ ions. Together with LEIS, two other techniques are used: Auger electron spectroscopy (AES) to analyze contaminants, having an analyzing depth of a few monolayers, and low-energy electron diffraction (LEED) to get structural data on both the specimen surface and probable superstructures on the surface. The cleaning procedures have been chosen for the SA specimen W(1 1 0) as the most characteristic case. Experimental measurements by LEIS, AES, and LEED before and after cleaning operations are presented in **Figures 3–5**. LEIS shows that W is nearly absent in the outermost atomic layer, showing that the specimen is almost completely covered with contaminants (**Figure 3**, untreated). Auger analysis also shows that C and O alone cover the surface (**Figure 4**, untreated). A significant reduction in C and O is received by sputtering the surface with 3 keV Ar^+ at RT. Notwithstanding, after sputtering the LEIS spectrum gives a pure W peak, and the contaminants are not completely extracted: in fact, sputter cleaning removes solely the contaminants of the outermost layer(s).

The high-temperature heating of the specimens restores their sputter-induced surface chaos and stimulates the C migration out of the undersurface location to the outermost layer. An optimal way of a C extraction from the outermost layer, together with the C depletion from an undersurface location, consists in heating the W specimens in O_2 gas. As a rule, the C depletion is made in a separate chamber by annealing the specimens with EB at 1500 K in O_2 gas (10^{-5} mbar). A release of C from the outermost layer stimulates the further traveling of C to the surface, which is afterward released by vacuum extraction of CO. At the final steps, the specimens have been flashed in UHV at 2500 K to extract O from the surface. C and O remaining after this procedure are removed by repeated cleaning. As a result of the C migration from the volume to the surface layers at the initial steps, a superstructure on the W(1 1 0) specimen forms, which is shown in **Figure 5**. After continued treatment the surface region is completely free of C, and alone O is present on the W surface after annealing in O_2. So, no more C migrates to the surface of the specimens when the surplus O is extracted by flashing.

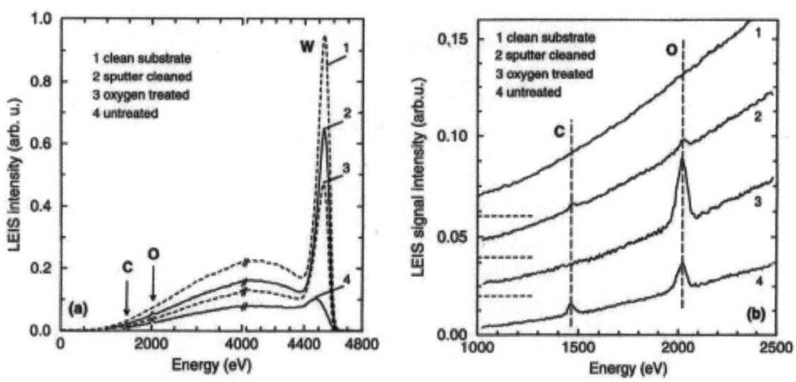

Figure 3.
LEIS 5 keV He⁺ spectra obtained after different cleaning steps of the W(1 1 0) specimens: (a) complete spectra and (b) energy range covering only the C and O peaks. The baselines of the spectra in (b) are indicated by the dashed lines. The AES spectra if the corresponding stages are shown in ***Figure 4****.*

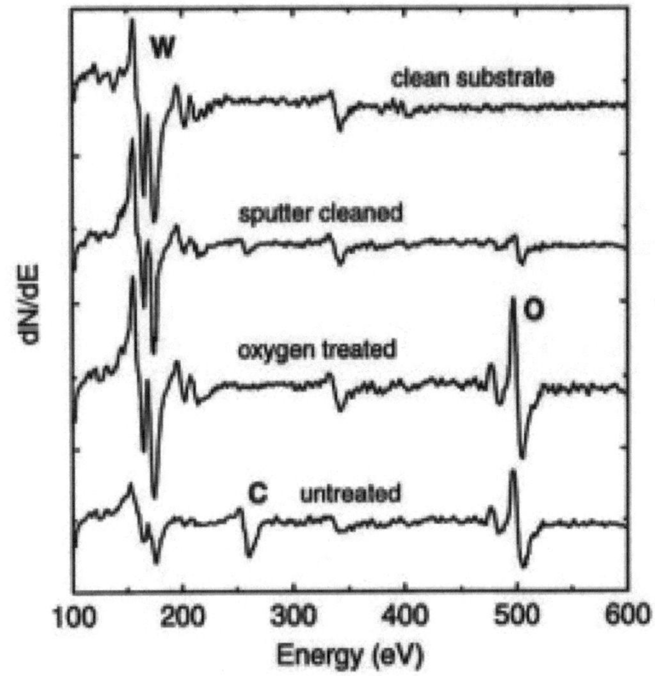

Figure 4.
The AES spectra at the different cleaning stages of the W(1 1 0) specimens. The LEIS spectra of the corresponding stages are shown in ***Figure 3****.*

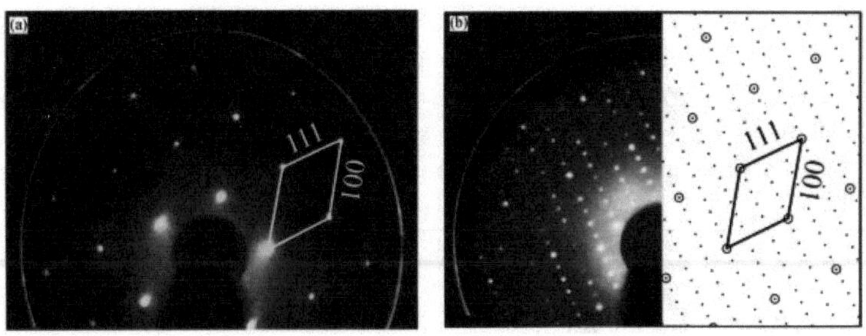

Figure 5.
LEED patterns of the W(1 1 0) specimens at an energy of 300 eV: (a) clean unreconstructed W(1 1 0) and (b) C superstructure after flashing in initial cleaning.

2.4 Tensile samples

Tensile specimens are produced by a step-by-step treatment. At first, to get small rods with 1.6 mm in diameter and 27 mm in length, large AG crystals are cut [8–12]. Small rods are processed by a round hollow electrode to decrease in diameter of a gauge part. After processing, the specimens of 1.3 mm in diameter and of 14 mm in length are produced, with shoulders of 1.6 mm in diameter on both ends. Then, the specimens are chemically polished to move away a damaged layer of 0.05 mm. Then, the specimens are suspended in the center of the hollow electrode. Both the electrode and specimen are put into 1% NaOH which flows through the hollow electrode along the specimen. When an electric current is on, the diameter of a specimen is decreased by etching to its final sizes of 0.9 mm and free from of

0.15 mm of a damaged layer. Such controllable simple step-by-step treatment allows to get a constant diameter along the gauge part. The tensile specimens are plastically deformed at a constant strain rate of 8.5×10^{-4} s^{-1} on two deformation setups with different *T*. Below, 320 K tests are done in a He cryostat which is fixed on an Instron. Type-A tests consist in isothermal plastic deformation of several samples to their ultimate tensile strain at different *T*. Type-B tests represent successive deformation in a little strain interspace at different *T* after a first deformation at >600 K.

2.5 Experimental results

In **Figure 6**, the shear stress τ is depended on the shear strain ε_p for samples deformed to neck formation or fracture at constant *T*. Parabolicity of the curves below 400 K in **Figure 6** connects with severe hardening at little ε_p. At T = 650 K, a slight hardening is seen up to $\varepsilon_p \leq 0.04$; however, at $\varepsilon_p \geq 0.04$, constant values of shear stresses are observed.

In spite of that, tensile specimens are produced with a great care; some of them can only be deformed to little ε_p, more pronounced at the lowest *T*. For ε_p = 0.001, the hardening rates increase more with lowering *T*, for example, to 70,000 MPa at 26 K. For $\varepsilon_p \geq 0.01$, the hardening rates are less at low *T* and a dependence of *T* lowers faster. In turn, parabolicity of the hardening curves prevents determining the critical shear stress. It is interesting that any problems do not appear if the specimens are subjected to small deformation at decreasing *T* after the prime

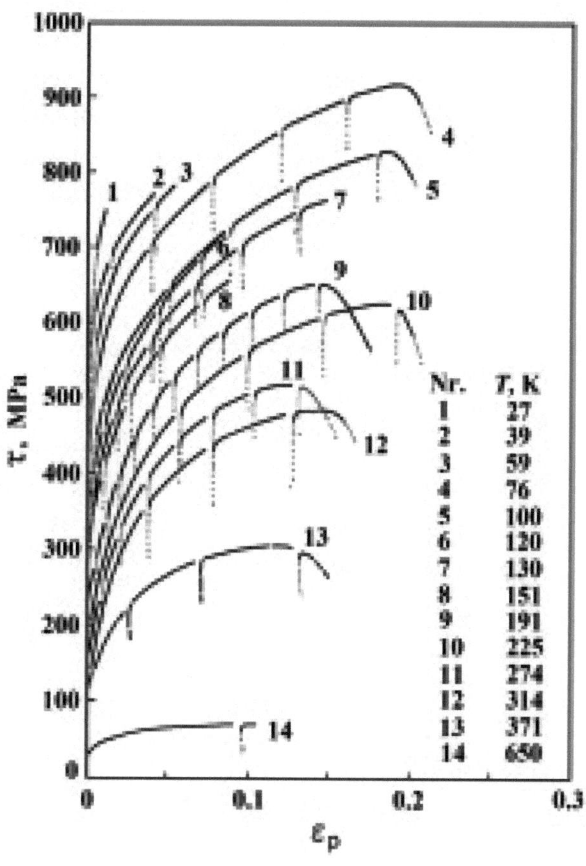

Figure 6.
Tests (A): shear stress τ versus shear strain ε_p of W at different T.

deformation to $\varepsilon_p \approx 0.08$ at $T > 600$ K (B-tests). **Figure 7** shows such curves received for one specimen for the shear stress τ as a function of the shear strain ε_p. It is important to note that at the prime deformation at 621 K after a short hardening regime, the stress has reached a *plateau*. However, subsequent plastic deformation at 586 and 541 K is not accompanied by hardening. At lower temperatures, the phenomenon of transition to the yield point (186 and 151 K) occurs.

In **Figure 8**, all experimental data of our tests are shown. The shear stresses, corresponding to the plateau and the lower yield stresses, are referred as a critical shear stress τ_c. They are plotted against T in the range of 50–800 K (**Figure 8**). The results of tensile tests in various cooling or heating conditions match in the best way [8–12]. Above 740 K, with increasing T, τ changes very little. The critical shear stress τ_c at 800 K is not high ($\tau \approx 13$ MPa), which is consistent with both the high purity and structural quality of tested W specimens. The value of τ down to low T increases very fast and parabolically. At 600–620 K, a pronounced transition to a semi-linear increase in τ is seen at lowering T to 200–250 K. Then, a plain transition to a solid rise in τ with lowering T can be seen. In accordance with this form of $\tau(T)$, it can be summarized that W single crystals of high-purity exhibit a three-stage

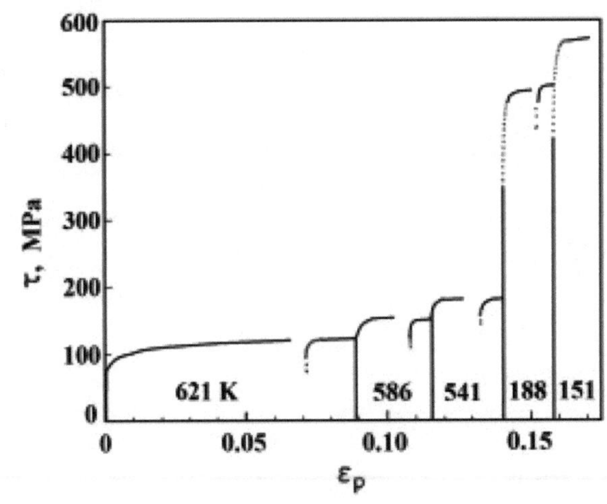

Figure 7.
B-tests: successive deformations of one W specimen at decreasing T.

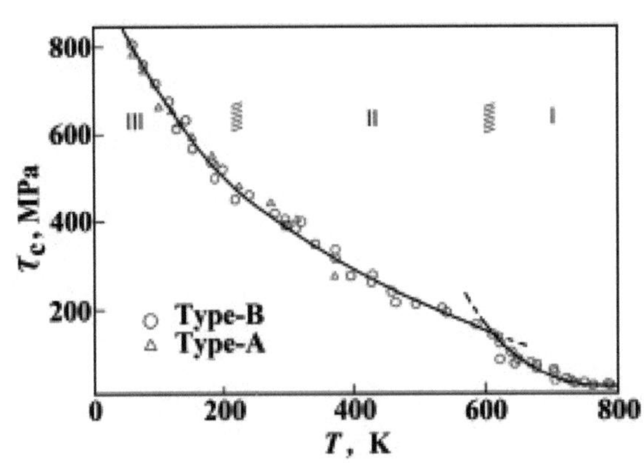

Figure 8.
The temperature dependence of τ_c for B-tests (A-tests also involved).

mechanical behavior (I, II, and III). This is also characteristic of plasticity of other BCC metals like Mo or Nb [13–17].

3. Conclusions

1. The estimates show that the axial gradients of T under the crystallization front in the solid phase can reach 1500 K cm^{-1}. This leads to significant thermal stresses followed by their relaxation through plastic flow and multiplication of dislocations. In other words, the temperature gradients cause high values of the dislocation density, which leads to the appearance of a characteristic dislocation substructure. In the crystals with the growth axis [1 0 0], there are small-angle boundaries elongated along the growth axis and having misorientations up to 1–2°. A fairly high ratio of the resistances of the investigated single crystals of W at the level of $\sim 10^4$ to 10^6 indicates a high purity of the metal, which leads to an unambiguous conclusion about the insignificant role of impurities in the development of the crystal dislocation structure.

2. W single crystals oriented for single slip have been tested in dynamical tensile tests between 26 and 800 K. The critical shear stress τ_c at 800 K is not high ($\tau \approx 13$ MPa), which is consistent with both the high purity and structural quality of tested W specimens. The value of τ down to low T increases very fast and parabolically. Qualitatively, these results agree well with type-A and type-B tests with high-purity Mo single crystals. The measurements confirm also for the W single crystals the existence of three regimes of the dependence of the flow stress on temperature which results meet quite well with known results of other BCC metals studied.

Author details

Vadim Glebovsky
Institute of Solid State Physics, The Russian Academy of Sciences, Chernogolovka, Russia

*Address all correspondence to: glebovs@issp.ac.ru

References

[1] Lassner E, Schubert WD, editors. Tungsten: Properties, Chemistry, Technology of the Element, Alloys and Chemical Compounds. New York: Kluwer Academic/Plenum Publishers; 1998

[2] Chaika A, Orlova N, Semenov V, Postnova E, Krasnikov S, Lazarev M, et al. Fabrication of [001]-oriented tungsten tips for high resolution scanning tunneling microscopy. Scientific Reports. 2014;**4**:3742

[3] Glebovsky V. Highly pure refractory metals for thin film metallization of VLSI. Chapter 13. In: Ho Yeap K, editor. Very-Large-Scale Integration. Rijeka: InTech; 2018. pp. 67-109. ISBN: 978-953-51-5309-2. DOI: 10.5772/intechopen.69126

[4] Cortenraad R, Ermolov S, Semenov V, Denier van der Gon A, Glebovsky V, Bozhko S, et al. Growth, characterization and surface cleaning procedures for high-purity tungsten single crystals. Journal of Crystal Growth. 2001;**222**:154-162

[5] Bozhko S, Glebovsky V, Semenov V, Smirnova I. Study on the growth of tungsten single crystals of high structural quality. Journal of Crystal Growth. 2008;**311**:1-6. DOI: 10.1063/1.3254240

[6] Glebovsky V. In: Sztwiertnia K, editor. Crystal Growth: Substructure and Recrystallization. Rijeka: InTech; 2012. pp. 59-86

[7] Cortenraad R, Ermolov S, Semenov V, Denier van der Gon A, Glebovsky V, Bozhko S, et al. Electron-beam growing and purification of W crystals. Vacuum. 2001;**62**:181-188

[8] Brunner D, Diehl J, Glebovsky V. The plastic behavior of high-purity tungsten. In: Proceedings of the 5th International Conference on Ultra High-Purity Materials (UHPM-98); Sevrier, Annecy Lake, France; 1998. pp. 83-92

[9] Brunner D, Glebovsky V. The plastic properties of high-purity W single crystals. Materials Letters. 2000;**44**: 144-152

[10] Brunner D, Glebovsky V. Analysis of flow-stress measurements of high-purity W single crystals. Materials Letters. 2000;**44**:290-296

[11] Brunner D. Comparison of flow-stress measurements on high-purity tungsten single crystals with the kink-pair theory. Materials Transactions of Japan Institute of Metals. 2000;**41**: 152-160

[12] Brunner D. Peculiarities of work-hardening of high-purity tungsten single crystals below 800 K. Materials Science and Engineering A. 2004;**387**: 167-170

[13] Hull D, Beardmore P, Valentine A. Crack propagation in single crystals of tungsten. Philosophical Magazine. 1965; **12**:1021-1041

[14] Butler B, Paramore J, Ligda J, Ren C, Fang Z, Middlemas S, et al. Mechanisms of deformation and ductility in tungsten —A review. International Journal of Refractory Metals and Hard Materials. 2018;**75**:248-261. DOI: 10.1016/j.ijrmhm.2018.04.021

[15] Wang J, Zeng Z, Weinberg C, Zhang Z, Zhu T, Mao S. *In situ* atomic scale observation of twinning-dominated deformation in nanoscale body-centred cubic tungsten. Nature Materials. 2015; **14**(6):594-600. DOI: 10.1038/NMAT4228

[16] Liu G, Song, Liu M, X, Ni S, Wang S, He Y, Liu Y. An investigation of the

mechanical behaviors of micro-sized tungsten whiskers using nanoindentation. Materials Science and Engineering A. 2014;**594**:278-286. DOI: 10.1016/j.msea.2013.11.084

[17] Riedle J, Fischmeister H, Gumbsch P, Glebovsky V, Semenov V. Fracture studies of tungsten single crystals. Materials Letters. 1994;**20**:311-320

Section 2

Growing Single Crystals

Chapter 2

Doping of SiC Crystals during Sublimation Growth and Diffusion

Evgeniy N. Mokhov

Abstract

The preparation of SiC crystals doped with various impurities introduced during the process of sublimation growth and diffusion is described. The growth of SiC crystals was carried out by a sublimation-sandwich method, proposed by us in 1970. Crystals of the n- and p-type conductivity with maximum content of electrically active impurities (of the order of 10^{21} cm^{-3}) were obtained. The solubility values of more than 15 impurities were determined. Special tantalum containers with several temperature zones, allowing to introduce any impurity into SiC practically, are developed. The dependences of the impurities concentration on the temperature, growth rate and seed orientation are found. Diffusion of impurities of boron, aluminum, gallium, beryllium, lithium, nitrogen and phosphorus in silicon carbide polytypes was studied. Diffusion coefficients of these impurities in a wide temperature range are determined. It is shown that for a number of impurities diffusion cannot be described by standard distributions that are explained by the presence of several states of diffusing centers. Fast-diffusing states are atoms located in interstices, as well as centers, including the impurity atom and point defect. The extremely low diffusion mobility of lattice point atoms in the SIC lattice is noted.

Keywords: doping, impurity, diffusion, sublimation growth, polytypes

1. Introduction

Controlled doping of the semiconductors is a necessary condition for the creation of device structures on their basis. The main attention in the SiC doping is usually paid to impurities with shallow levels in the bandgap—N, Al and B [1, 2]. These impurities are characterized by high solubility in SiC that allows to receive low-resistance SiC layers [3]. Other impurities—elements of II, III, IV and V groups, transitional metals and rare-earth elements—are less often used for SiC doping, though they are of interest for fabrication of high-resistance and semi-insulating SiC layers and special types of light-emitting diodes. The impurities may effect on the polytype structure [4]. For example elements of IV group allow to grow the 4H-SiC crystals on the seeds of other polytypes. Finally, incorporation of impurities is necessary for identification of the nature of impurity centers, which is important for prospects assessment of creation of new semiconductor devices generation.

Doping of SiC crystals is carried out by various methods: during the growth process, by ion implantation and by thermal diffusion. A perspective way is also transmutation doping. It allows to receive SiC crystals uniformly doped with impurities. In this review we will limit ourselves to the description of SiC doping during the sublimation growth and by thermal diffusion.

The doping of SiC is characterized by a number of specific features. Doping process is carried out at high temperatures (1700°C and higher); concentration of doped impurities considerably depends on various factors, such as growing surface conditions (crystallographic orientation, density of structural and morphological defects, and thickness of the growth steps [5]), stoichiometry deviation and structure of the polytypes [6, 7]. Incorporated impurities easily form precipitates [6] and durable, thermal stable associates with matrix atoms or native defects [8].

The goal of the present work is observation of the SiC doping by different impurities during the sublimation growth or the thermal diffusion. Impurities of II, III and IV groups of the periodic table, the transitional metals and rare-earth elements were incorporated into SiC crystals of various polytypes. The conducted investigations have allowed not only to define solubility of these elements in SiC but also to identify the impurity centers and to estimate their influence on semiconductor SiC properties. Some results of this work were presented previously [8–11].

2. SiC doping during sublimation growth

Impurity incorporation was carried out during physical vapor transport (PVT) growth. Usually SiC crystals or epitaxial layers were grown by sublimation sandwich method (between the source and the seed (no more than 0.2 from the linear size of the source). It is important that at a small gap in the growth cell a self-contained system is realized. It provides quasi equilibrium conditions on the growing crystal surface [12]. The schematic diagram of SSM is shown in **Figure 1**. The feature of SSM is the existence of a narrow gap and the evaporated surfaces in the wide range of temperatures and pressures. As a result the temperature range of SiC crystal growth by SSM is very wide: from 1700 to 2700°C. Growth of high-quality crystals and epitaxial layers (EL) is possible in vacuum and in gas phase environment. SSM is successfully used for growth of SiC bulk crystals [13] and epitaxial layers [14] and also growth of GaN [15] and AlN [16] crystals.

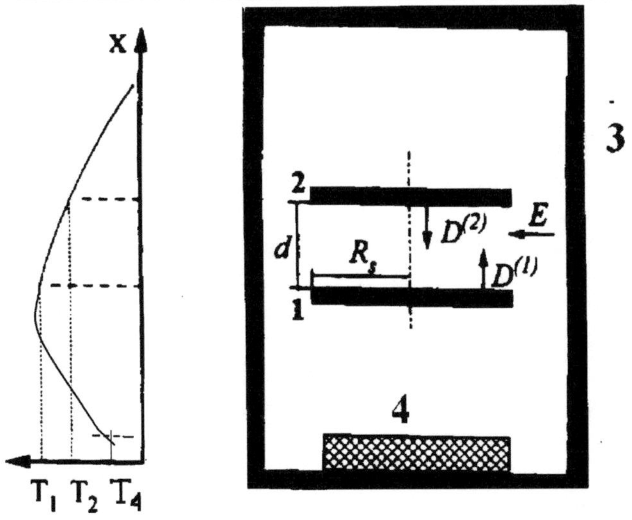

Figure 1.
Schematic view of the sublimation sandwich system for growth of the doped SiC crystals. Source of SiC vapor (1); seed (2); container (3) and source of impurity (4). Temperature distribution is shown (left).

In this method high efficiency of impurity mass transfer is reached that is difficult to realize in other gas transport techniques, for example CVD [17].

For SSM the influence of impurities on the quality of the growing crystal is not so essential. Therefore, SSM is an effective method for growth of the high-doped SiC crystals with solubility limit of the impurities. The choice of impurity is almost unlimited.

The schematic diagram of the container used by doping is shown in **Figure 1**. By growth of doped crystals in a crucible there are no less than two temperature zones. The source of SiC vapors and the seed are placed into the hot zone. The impurity is located in the zone with lower temperature. The concentration of the entered impurity is defined by impurity source temperature. For growing of heavily doped crystals, it is preferable to use the impurity source, which is in an elementary state. The impurity can also be loaded directly in SiC source. Volatile impurities (for example, nitrogen) are entered through the gas phase.

Growth is usually carried out in graphite or tantalum containers. The use of a Ta container helps to reduce the content of background impurities, to reduce sharply the loss of the sublimated material and to provide growth of SiC crystals without graphitization of the source [18]. The last is very important for receiving high-quality SiC crystals with low dislocation and micropipe density. Unfortunately, Ta impurity creates a deep level in band gap [19].

For determination of impurity content in SiC crystals neutron-activation analysis [20, 21] secondary ion mass-spectroscopy (SIMS) and different electro-physical methods (van der Pauw, surface local breakdown [22], and C-V measurements) were used.

2.1 Solubility limit of the impurities in SiC crystals

Solubility limit of more than 20 impurities incorporated during PVT growth by SSM is presented in **Table 1**. Besides, we studied doping of SiC crystals with Fe, Ni, and Er impurities and found that its solubility in SiC was at the level of 10^{17} cm^{-3}. Neutron-activation analysis (NAA) [20] was used for determination of the impurities concentration in uniformly doped parts of crystals.

As shown in **Table 1** there is very limited set of impurities with high solubility in SiC. These are, first of all, acceptors (Al, B, Be, and Ga), donors (N and P) and Ge. For the majority of impurities the maximum level of SiC doping is reached at extremely high temperatures (Tg > 2400°C) and for [0001]Si growth direction.

Element	Concentration cm^{-3}	Element	Concentration cm^{-3}	Element	Concentration cm^{-3}
Li	1.2×10^{18}	Cr	3.0×10^{17}	In	9.2×10^{16}
Be	8.0×10^{20}	Mn	3.0×10^{17}	Sn	10×10^{16}
B	2.5×10^{20}	Cu	1.2×10^{17}	Sb	8.0×10^{15}
N	$8.0 \times 1\,0^{20}$	Ga	1.8×10^{19}	Ho	6.0×10^{16}
A1	2.0×10^{21}	Ge	3.0×10^{20}	Ta	2.4×10^{17}
P	2.8×10^{18}	As	5.0×10^{16}	W	2.5×10^{17}
Sc	3.2×10^{17}	Y	2.0×10^{16}	Au	4.9×10^{16}
Ti	3.3×10^{17}	Mo	3.5×10^{17}		

Table 1.
Solubility limit of impurities in SiC crystals doped during sublimation growth [23].

2.2 Dependence of doping of impurities on crystallographic orientation

It was determined that SiC doping level strongly depends on crystallographic orientation of the substrate. Higher concentration of impurities of the III-a group (In, Al, Ga) and also transitional elements is observed by growth in the direction of [0001]Si, in comparison with the direction [0001]C [24–26] (**Figure 2**).

On the contrary the concentration of V-group impurities (N, P, As and Sb) is higher by growth on the [0001]C surface [23, 26, 27]. At low growth temperatures the effect of orientation anisotropy of SiC doping is very considerable. For example the concentration of acceptor Al and Ga impurities grown on polar {0001} sides at the temperature 1800°C differs 5–10 times [24, 25]. The dependence of impurity concentration on substrate orientation noted above also remains by inversion of the sample conductivity type.

The obtained data are explained by the absence of equilibrium vapor-crystal at the typical growth conditions. It is known [27] that the condition of equilibrium vapor-crystal is the inequality: $V_g < D_i/h$ (V_g—growth rate, D_i—impurity diffusion coefficient, h—thickness of the growing layer).

We have obtained that for impurities N, P, Ga and Al [9] at practically realized temperatures and growth rates this condition is not satisfied and doping anisotropy is a consequence of various adsorptive properties of polar {0001} sides.

It is known that chemical bonds in the surface layer are rehybridized in such a way that individual properties of the element located on the surface become essential [28]. This feature results in differences in the character of surface sides, which consist of only silicon or carbon atoms. With the increase of temperature the

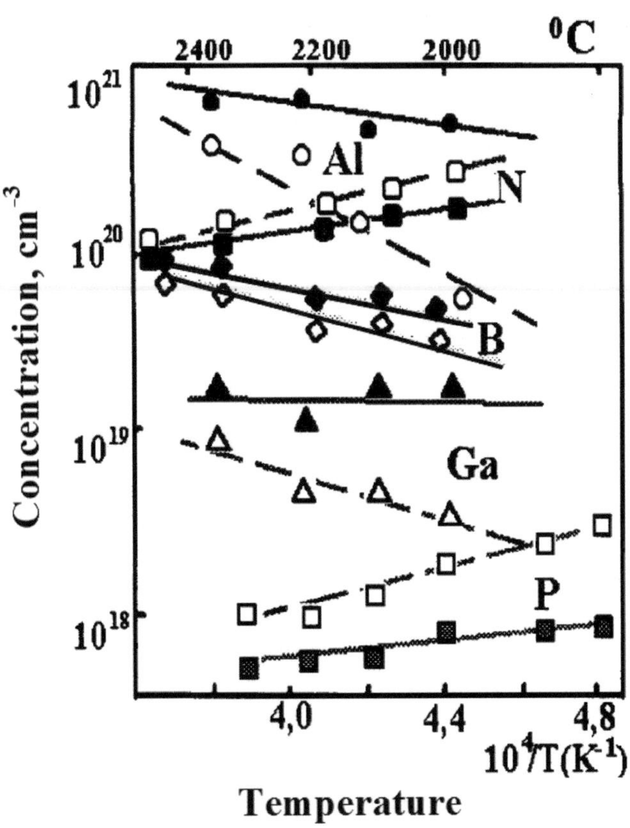

Figure 2.
Dependence of the concentrations of N, B, Ga, Al, and P in SiC on the growth temperature. Orientation of the growing SiC: [0001]Si, solid lines; [0001]C, dashed lines.

anisotropy of polar sides doping decreases. A similar effect is got by introduction of silicon vapors [29, 30] or impurities, such as tantalum, promoting enrichment of the surface layer by the silicon into the growth cell [31].

The influence of substrate orientation also becomes apparent at small angles of its misorientation in relation to singular {0001} planes [32]. For example nitrogen concentration in epilayers grown on on-axis (0001)C face is 1.7–2 times higher than on off-axis one. This results from the fact that increase of the growth steps density in the second case leads to a raised desorption of impurity atoms. Such strong dependence of the doping impurity concentration on the misorientation angle leads to nonuniform doping of the grown crystal and emergence of morphological imperfections on the growing surface [32]. Impurities incorporation also depends on growth layer thickness due to step bunching process [33].

As a rule, with temperature increases, the concentration of impurities is enhanced (**Figure 2**). However, for V-group elements nitrogen and phosphorus the opposite effect was observed, i.e., the doping level went down with the temperature (**Figure 2**). The abnormal temperature dependence for nitrogen and phosphorus is explained by the fact that capture of these impurities is limited by the process of desorption in which probability increases with temperature rise [34].

Thus, for receiving SiC samples heavily doped by acceptor impurities high temperatures of growth are preferable (2400–2500°C) and low-resistance layers of n-type conductivity with extremely high content concentration of nitrogen or phosphorus can be grown at the moderate temperatures (1800–1900°C).

In all cases the dependence of $C_i(T)$ can be described by Arrhenius's equation: $C_i = A \exp(-\Delta H/RT)$, where ΔH is dissolution enthalpy. If the doping impurities are entered in the process of crystal growth, values A and ΔH strongly depend on crystallographic orientation of the substrate. For IIIa-group impurities the values A and ΔH are significantly higher for growth in the direction of [0001]C than in the direction of [0001]Si [35].

2.3 Influence of partial pressure of the impurity

Dependence of impurities concentration of Al, B, Ga and N on partial pressure is presented in **Figure 3**. Conditions of saturation achievement differ strongly for each impurity. It is seen (**Figure 3**) that in case of B and Al impurities the solubility limit is now realized by rather low partial pressure of impurity (about 10^0–10^2 Pa) [23]. Solubility limit of Ga impurity is observed at pressures 10^2–10^3 Pa [36]. But the saturation of N concentration even at extremely high pressures of molecular nitrogen is not observed [37].

The dependence $N_d \sim P_i^{1/2}$ is realized in very wide range of pressures (10^0–10^5 Pa) (where N_d—concentration of donors, P_i—partial pressure of nitrogen). It is a consequence of the Henry's law implementation and assumes existence of equilibrium like $N_2 \leftrightarrow 2N$. At the same time the density of adsorption centers on the surface is much higher than impurity concentration in chemosorbate because of very small life time of nitrogen molecules on the growth surface [37].

2.4 Dependence of the doping level on the growth rate

Growth rate in some cases also influences concentration of the entered impurity. We observed the diminishing of N impurity concentration at high growth rates (**Figure 4**).

In these experiments the impurity source was combined with the SiC source and located inside the growth cell [30, 31]. The influence of the growth rate on the level of doping by nitrogen has been quantitatively explained within the model according to which the capture of impurity is limited by kinetics of adsorption [31].

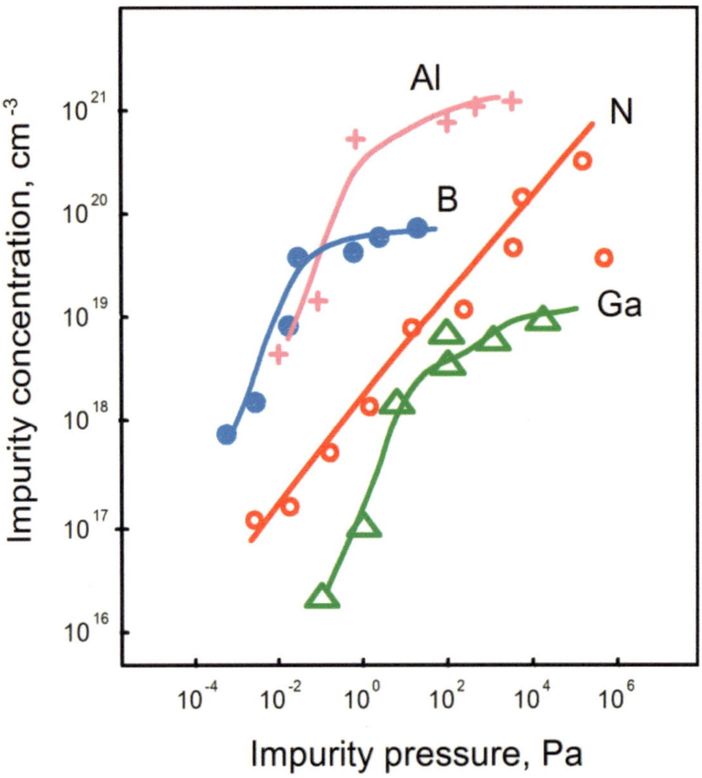

Figure 3.
Dependence of the impurity concentration in the growing SiC on the dopant partial pressure inside the growth cell for the cases of doping by N, B, Al and Ga. T = 2000°C. Doping with N is performed on (0001)C surface; with B, Al and Ga on [0001]Si surface.

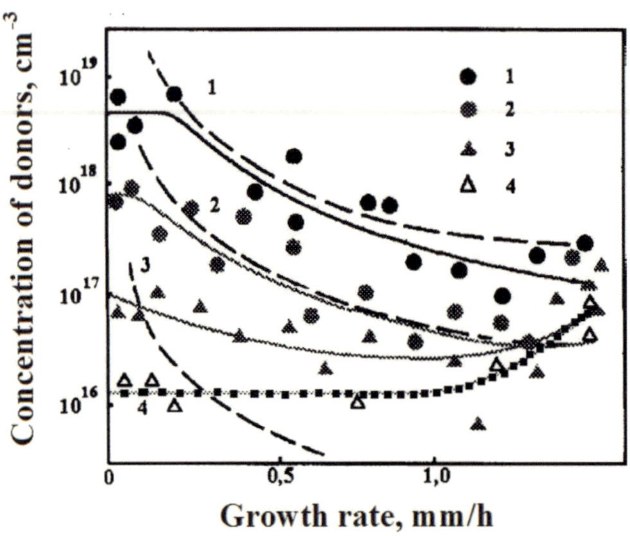

Figure 4.
Dependence of the donor concentration (N) in SiC epitaxial layer on growth rate. Concentration of nitrogen in the source: (1) 2 × 10^{19}, (2) 3 × 10^{18}, (3) 3 × 10^{17} and (4) 7 × 10^{16} cm^{-3}. The growth temperature is 1850°C. Dashed lines are theoretical results.

However, good correlation data calculation with the experiment was observed only when the heavily doped nitrogen SiC source was used [31]. At low concentration of nitrogen in the source ($<10^{18}$ cm^{-3}) the concentration of donors in the grown crystal either poorly depended on the deposition rate or even increased with its increase. The obtained result has been explained by the fact that by growth of SiC there could take place not only capture of impurity (nitrogen) atoms but also nonequilibrium native defects of donor type. Their concentration increased with the increase of the deposition rate. Native defects in SiC crystals grown at high rates were found also by other researchers.

On the contrary increase of growth rate leads to higher Ga impurity incorporation [35]. At low growth rates (Vg < 10 μm/h) the concentration of Ga in SiC did not exceed (3–5) × 10^{17} cm^{-3}. The increase of growth rate up to 0.2–0.5 mm/h leads to increase of impurity concentration up to 10^{19} cm^{-3}. This dependence can be explained within the conception of nonequilibrium capture of impurity [27].

2.5 Influence of stoichiometry deviation on the doping level

As it is stated, the concentration of Al, Ga and N impurities in the grown SiC layers significantly depends on the ratio Si: C in the vapor phase [32]. By introduction of silicon vapors into the growth zone the doping level of Al and Ga impurities considerably decreases. Especially, sharp decrease of concentration of these impurities (almost in order of magnitude) is observed by growth in the direction of [0001] Si [32]. As a result it appears impossible to receive low-resistance layers of p-type conductivity in the SiC-Si system. By surplus of silicon the concentration of nitrogen in the grown SiC layers also decreases [29]. A similar dependence of impurity concentration on surplus of Si is revealed for growth of SiC layers by the method of gas transport deposition [38]. Obviously, the reduction of impurity capture efficiency by growth in the SiC-Si system is a consequence of the competing adsorption of silicon [32]. Therefore, density of sorption centers on the growing surface decreases.

Besides, it was found that introduction of silicon vapors to the growth zone leads to sharp reduction of orientation anisotropy of doping SiC layers by N, Al and Ga impurities [26, 32]. It facilitates receiving of bulk SiC crystals with more uniform impurity distribution.

We note that a similar effect of the influence of stoichiometry deviation, known as site-competition epitaxy, is described in the works of [38]. By growth of SiC crystals by CVD method the authors found that high tension of Si vapors promotes decrease of the acceptor doping level. On the contrary, at excess of carbon, the concentration of nitrogen decreases, especially by growth in the direction of [0001]Si. The concentration of electrically active impurity depends on Si/C ratio in the gas phase (CVD). High concentration of C prevents introduction of nitrogen atoms but only on the Si side (decrease by 4–5 times). It has allowed to receive crystals with (Nd-Na) = 10^{14} cm^{-3} [38]. For acceptor impurities excess of Si lowers their concentration.

2.6 Coefficients of impurity capture

Within the bounds of classic approach for quantity description of doping process coefficient K_t of impurity transfer from source to substrate is usually used. But K_t depends on impurity contents in the source, on geometrical sizes of the source and the substrate and on the distance between them.

The value defining probability of the impurity capture by a single collision with growing surface is the coefficient of elementary impurity capture (K_i). It is easier to

Impurity	K_i, [0001]Si	K_i, [0001]C	Ref.
N	5.0×10^{-2}	1.2×10^{-1}	[35]
B	4×10^{-2}	2×10^{-2}	[35]
A1	8×10^{-3}	4.5×10^{-4}	[35]
P	2×10^{-4}	5×10^{-3}	[40]
Ga	1.5×10^{-3}	4.5×10^{-5}	[35]

Table 2.
Impurity unit capture coefficients (K_i) for sublimation SiC growth at the temperature 1850°C and growth directions [0001]C and [0001]Si.

define K_i value by transfer of impurity from the source to the substrate in the conditions of vacuum [34, 39]. For this purpose SiC bulk crystals with a certain content of doping impurity are usually used as a source.

By means of this technique we have calculated the unit impurity capture coefficients (K_i) for the most important electrically active impurities—nitrogen, boron, aluminum, gallium, phosphorus—depending on the growth temperature, substrate orientation and impurity concentration [35, 40].

Some results of this research are shown in **Table 2**.

It is clear that during sublimation growth the most effectively captured impurities are nitrogen and boron. Unit capture coefficients of other impurities are much less.

The main reasons of low coefficients of impurity capture during sublimation growth are high probability of desorption of impurity atoms and their interaction with matrix atoms with formation of precipitates.

At high concentration of impurity essential decrease of K_i value is observed (**Figure 5**) [23, 37, 40]. In case of nitrogen and phosphorus this effect can be explained by the fact that with impurity concentration increase in the adsorbing layer, probability of its desorption in the form N or P molecules increases. Reduction of Ki boron and aluminum is possibly caused by limitation of centers of effective sorption or is a consequence of these impurities' precipitates formation.

2.7 Macrosegregation of impurities

In this work essential attention was paid to study the behavior of impurities entered into crystal. It has become clear that in heavily doped SiC layers, a considerable part of impurity is in inactive state and poorly influences material properties [41]. It has been found that the main reason for it is the formation of precipitates enriched by the doping impurities. Macrosegregation of impurities

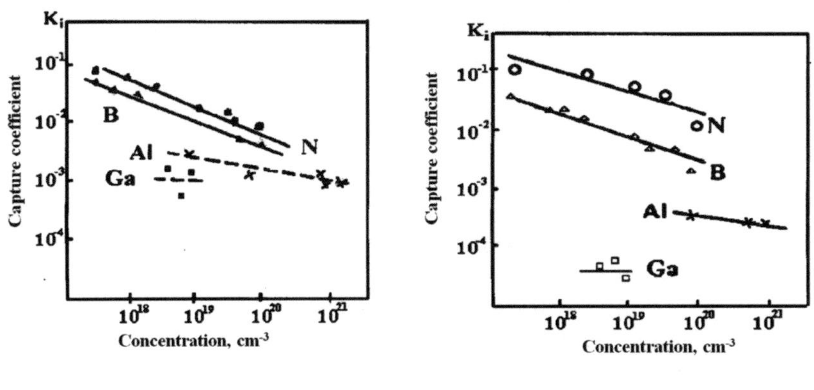

Figure 5.
Dependence of unit capture coefficients of impurities on the source impurity concentration. Growth directions: [0001]Si (left) and [0001]C (right).

caused by formation of precipitates leads to sharp deterioration of material and device structures on its basis. Therefore, we have paid much attention to this question.

In this work we have studied the reasons of precipitates formation. For this purpose we have investigated characteristic features of impurity macrosegregation in SiC layers grown from vapor phase depending on temperature, growth rate, substrate orientation and vapor phase structure [42–44]. The behavior of various impurities including Al, B, Ga, Ta and W in SiC crystal doped by growth and diffusion has been studied. Definition of impurity concentration and the nature of distribution were carried out by the combined method of neutron activation analysis including tool and autoradiographic options.

It was shown [34] that the precipitates enriched by impurities are usually observed in the SiC layers doped with Al, B, W, Ta and other impurities. As a result total concentration of these impurities in the doped layers can exceed on 2–3 orders of magnitude of the value of solubility limit typical for homogeneous solid solution. X-ray analysis shows that precipitates in SiC are usually carbides of the doping impurities.

Precipitates were observed in SiC crystals with low impurity content. For example precipitates enriched by boron impurity were discovered in SiC crystals, in which concentration of boron in homogeneous solid solution was near 10^{16} cm^{-3}, i.e., four orders of magnitude lower than solubility limit [23].

Impurity rich precipitates are concentrated on the crystal surface mainly near dislocations, pores, macrosteps, interpolytype boundaries and other structural and morphological defects (**Figure 6**) [36]. As a rule, higher impurity content caused by existence of precipitates was observed in SiC layers grown on (0001)C face. But concentration of the same impurities in homogeneous solid solution, on the contrary, is higher for layers grown on (0001)Si face. According to the results of autoradiographic investigations it is caused by larger sizes of the inclusions which are formed by growth on (0001)C faces [43].

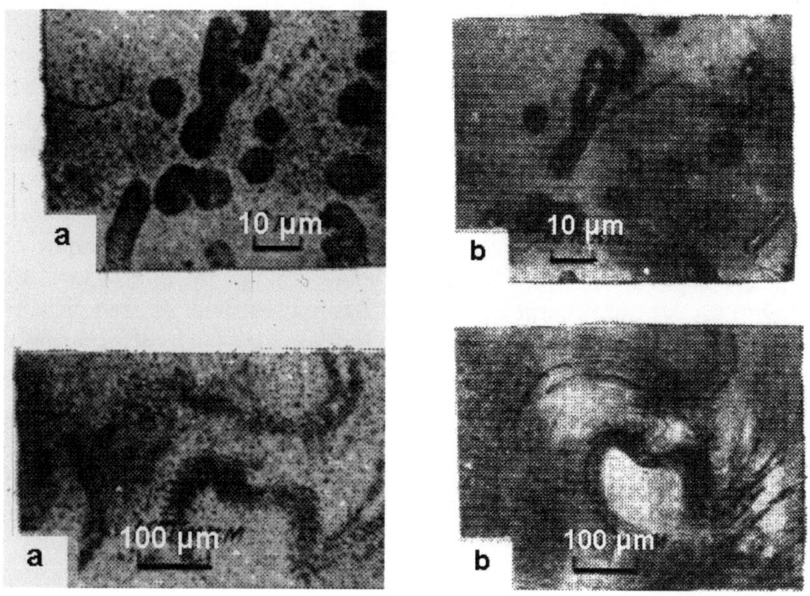

Figure 6.
Alpha-particle autoradiogram of 6H SiC sample after diffusion of boron from vapor phase (magnitude X40). It can be seen that second phase inclusions enriched boron are concentrated mainly at points of dislocations emergence and macrostep boundaries (a). Optical microscopy of the same parts of the surface is shown (b) [42].

At high growth rates the probability of precipitates enriched by impurities formation increases, which leads to worse crystal quality [43]. It is interesting that high density of precipitates leads to the decrease of impurity concentration in homogeneous solid solution. It was explained by the fact that precipitates could be getters for impurity atoms.

Features of macrosegregation on polar {0001} faces are explained by various surface energies of these sides. Lower surface energy of the C-face promotes formation of three-dimensional germs on it. Taking into account that formation of precipitates reduces desorption of impurity atoms, it becomes clear why total concentration of impurity is higher in the layers grown on the C-face than on the Si-face.

3. Doping of SiC during thermal diffusion

Diffusion of impurities is one of the main ways of device structures formation in semiconductors. Investigation of diffusion in solid states gives rich information on the nature of native defects, mechanisms of defect formation and migration and interactions of native defects with impurity atoms at high temperatures.

We have studied diffusion of impurities B, Al, Ga, Be, Li, N, P and O. Diffusion of impurities was carried out from vapor phase, and impurity vapor pressure was set by the source temperature. The investigated samples were located in the hot zone. In the colder zone a diffusant was placed. It was usually in an elementary state.

For measurement of diffusive distribution we applied the whole set of methods including special nuclear and physical techniques, which allowed to study the concentrational distribution of impurity both in intrinsic SiC and in heavily doped SiC samples. If these methods could not be used, the diffusive profile was defined with the help of standard semiconductor techniques, such as Hall method, C-V measurements and p-n junction method.

3.1 Diffusion of lithium

Diffusion of Li was carried out from vapor phase at the temperatures 1250–2200°C [45]. The method of track autoradiography [45], based on registration by a solid-state track detector the Li (n, a) H nuclear reaction products, was applied for direct definition of Li concentration in the diffusive layer by radiation of samples with a flow of thermal neutrons.

The results of study of Li diffusion in SiC are presented in [45]. They demonstrate that diffusive distribution of Li in SiC can be described by standard erfc function. The dependence of diffusion coefficient (D_{Li}) on the doping level and the conductivity type of the studied samples was not revealed. The temperature dependence D_{Li} on the temperature is shown in **Figure 10**. High diffusive mobility of Li atoms in combination with rather low activation energy (ΔE = 1.7 эВ) leads to claim that Li in SiC, as well as in Si and Ge, diffuses by interstitial mechanism.

Diffusion of Li and hydrogen in SiC was also studied in works [47, 48]. Impurities were entered into the crystal by ionic implantation. It was shown that these impurities diffuse with rather high rate along interstates, and diffusion parameters rather well correlate with the data of the work [45].

3.2 Diffusion of beryllium

Diffusion of Be in SiC was carried out at the temperatures 1700–2250°C [49]. The p-n junction method was used for diffusion profile discover. It was supposed that Be impurity near p-n junction is completely ionized. At temperatures of diffusion above 1900°C the concentration profile had two clearly marked regions. Each of them could be described by one coefficient of diffusion.

Diffusion of Be in p-SiC <Al> was studied at temperatures 1300–2000°C by the methods of layer-by-layer measurement of conductivity and Hall effect [50]. After diffusion of Be the reduction of value (Na-Nd), concentration of holes and their mobility were observed. Since the mobility of free holes falls and conductivity value along the impurity zone grows after diffusion of Be, it is possible to assume that the value (Na-Nd) decreases due to compensation of acceptors by the incorporated donor impurity (in this case, Be). It helped to define the diffusive profile of Be in p-SiC.

Parameters of diffusion of Be and Li in p- SiC were rather similar. Therefore, we made a conclusion that diffusion of Be, as well as Li, in heavily doped p-SiC is carried out by an interstitial mechanism. This assumption is supported by results of the analysis of donor-acceptor interaction [51], according to which donor Be has a charge +2. A small size of Be^{+2} ion (0.31A) also promotes interstitial diffusion of Be. Apparently, in n-type SiC, the mechanism of diffusion is more complicated. And in this case, interstitial Be also quickly diffuses. But in p-type SiC quickly diffusing interstitial atoms are captured by traps (obviously, vacancies), which leads to diffusion slowing down. We believe that another diffusive-active state is the associate of Be atom with a vacancy. Such associates are found by electron paramagnetic resonance method in Be-doped SiC crystals [52]. Diffusion of Be in SiC from ion-implanted layer was studied in [53]. The results of this work correlate well with our date.

3.3 Diffusion of boron in SiC

Boron is the most important acceptor impurity and the activator of luminescence in SiC. Its introduction by diffusion is widely used for creation of various semiconductor devices. For better understanding of B diffusion mechanism, we carried out the research of B diffusion in pure and doped SiC at a wide variation of experimental conditions. The influence of temperature, structure of the vapor phase, boundary conditions and the initial condition of impurity on diffusion were specially studied.

In the majority of experiments diffusion was carried out from vapor phase in the temperature range 1500–2600°C. The diffusive profile of boron atoms was defined by the method of track autoradiography based on registration of α-particles according to nuclear reaction: B^{10}(n, a) Li^7 [41]. The use of boron isotope B^{10} as diffusant allowed to increase sensitivity of the method up to 5×10^{15} cm^{-3} with an accuracy ±15%. Distribution of boron acceptors was found additionally by Hall measurements at 77–1100 K temperature range, capacitive and p-n junction methods [54, 55]. The results of the diffusive profile measurements received by different methods [41, 54–56] rather well correlated with each other, except heavily doped near-surface layer, where concentration of boron atoms could exceed concentration of acceptors by 1.5–2.0 times [41].

As a rule concentration distributions of boron had a step two-branch profile and could not be described by one coefficient of diffusion (**Figure 4**) [55]. It has been stated that the character of diffusion distribution and rate of diffusion strongly depend on the value of surface concentration (C_s) [55]. For low Cs < 1×10^{18} cm^{-3} the diffusive profile could be described by one coefficient of diffusion. At increase of Cs the rate of boron diffusion in the volume region of distribution increased and the concentration profile took a step form. Besides, at high temperatures (>2100°C) on the "tail" of the volume branch at $C_V < 5\text{–}10 \times 10^{17}$ cm^{-3} a region with an abrupt inclination of the diffusive profile occurred. More detailed research of concentration distribution has allowed to find a minimum on the border of near-surface and volume regions [25]. For finding a charging condition of diffusion centers and identification of the impurity migration mechanism we carried out the diffusion study in SiC highly doped by donor and acceptor impurities.

3.3.1 Diffusion of boron in highly doped SiC crystals

We have studied diffusion of boron isotope B^{10} in SiC of n- and p-type conductivity doped with Al and N. Diffusion distribution B in SiC highly doped with nitrogen are presented in **Figure 7a**. The obtained results definitely show that nitrogen impurity leads to slowing down the rate of boron diffusion [57, 58]. This effect was especially considerable at high concentrations of nitrogen and low temperatures of diffusion, when the coefficient of boron diffusion decreased almost to an order of magnitude. It could not be explained only by the donor-acceptor interaction, as it took place, in particular, in intrinsic material when concentration of impurity is lower than n_i. Therefore, the main reason for slowing diffusion down may be the formation of complexes according to reaction: Bsi + Nc → (BN). This interaction is promoted by high BN binding energy, and the fact that these impurities replace different units in the SiC lattice.

On **Figure 7b** concentration profile of Boron in SiC doped with Al impurity is shown. Concentration of impurities in the SiC samples was changed from 10^{17} to 10^{21} cm^{-3} [57]. According to the obtained results (**Figure 7b**) by increase of acceptor concentration the rate of boron diffusion in SiC considerably increases. In case of low surface concentration $Cs < 1 \times 10^{18}$ cm^{-3}, when distribution in n-type, as well as in p- type, is described by one coefficient of diffusion, this dependence is well described by one coefficient of diffusion. It looks like: $D = D_i\,(p/n_i)$, where D and D_i—coefficients of boron diffusion in p-type and intrinsic SiC, correspondingly; p—concentration of free holes; and n_i—concentration of own carriers. The energy activation of diffusion in heavily doped SiC is 3.4 eV, that is 2.2 eV lower than in pure SiC. Thus, within the frames of vacancy model, it is possible to make a conclusion that diffusion of boron is carried out with participation of the vacancies with a charge +1.

3.3.2 Diffusion of impurities in SiC polytypes

Our research of diffusion of impurities in SiC crystals of various polytypes (3C, 4H, 6H, 8H, 15R, 21R and 27R) has revealed dependence of the diffusion rate

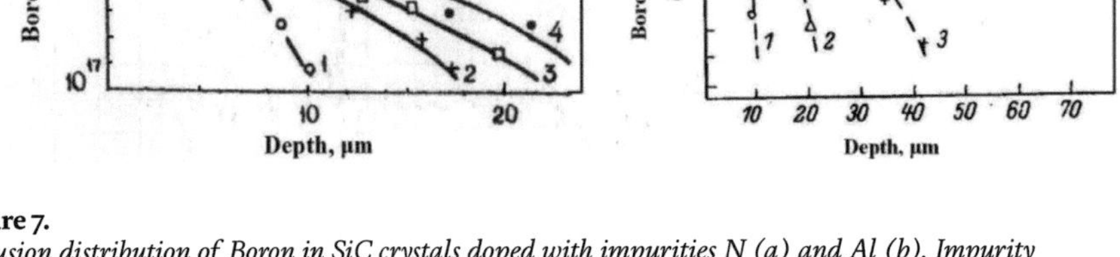

Figure 7.
Diffusion distribution of Boron in SiC crystals doped with impurities N (a) and Al (b). Impurity concentrations (cm^{-3}): N—1.1 × 10^{20} (1); 2-5 × 10^{19} (2); 1.5 × 10^{19} (3); 2 × 10^{17} (4). Al—4 × 10^{17} (1); 2 × 10^{18} (2); 6 × 10^{19} (3); 10^{20} (4).

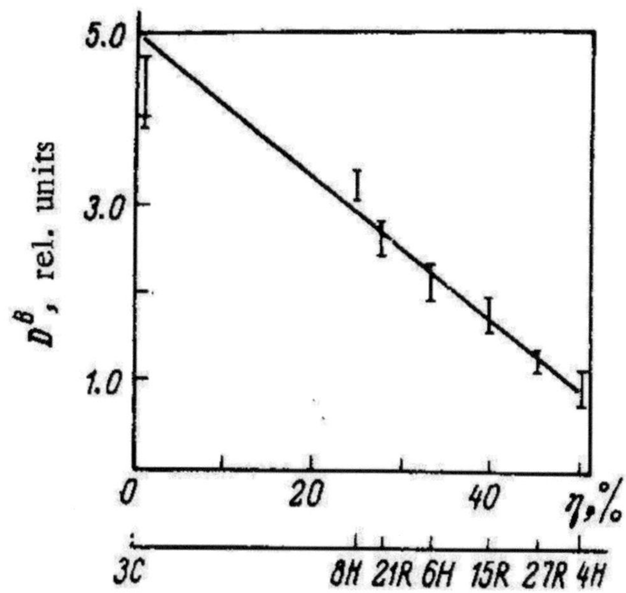

Figure 8.
Dependence of boron diffusion coefficient in SiC on polytype hexagonality percent.

on the SiC polytype (**Figure 7**) [59]. The results for boron diffusion are presented in **Figure 8**. They show the existence of a correlation between the value of diffusion coefficient and the degree of polytype hexagonality. The highest rate of diffusion is observed in cubic SiC. On the contrary, the lowest one is revealed in the most hexagonal 4H SiC polytype. Such dependence is caused by the increasing concentration of carbon vacancies in more cubic polytypes [4].

3.3.3 Constant-concentration diffusion of boron

For clarification of abnormal nature of diffusive distribution at high concentrations of boron, we have used a technique of constant-concentration diffusion [60]. There is practically no gradient of concentration of diffusing impurity in this case. For this purpose we studied diffusion of isotope B^{10} in SiC crystal preliminary highly doped with isotope B^{11} at the level of (5–6) × 10^{19} cm^{-3}.

According to the obtained data concentration distribution at constant-concentration diffusion of boron, unlike chemical diffusion, has a standard form (**Figure 9**).

By analysis it is necessary to consider that boron is an acceptor impurity in SiC and consequently, with increase in its concentration, provided that $C_B > n_i$, the coefficient of diffusion of boron should increase. However, the coefficient of constant-concentration diffusion of B was even lower than that of chemical diffusion in intrinsic SiC. The difference in diffusion mobility of boron becomes especially considerable if we compare the diffusion distributions of constant-concentration diffusion and boron distribution in p-type SiC doped with Al (**Figure 9**). The results of the experiment show that at identical concentration of acceptor impurity, the diffusion coefficient in p-SiC(Al) is nearly three orders of magnitude higher than in p-SiC (B).

The obtained results can be explained consistently, having assumed that boron at concentration higher than 10^{18} cm^{-3} creates traps for diffusive-active state of boron. Existence of these traps, apparently, is responsible for occurrence of abrupt near-surface region of the diffusive profile at usual chemical diffusion. It is natural to assume that such traps are boron atoms located in units of the lattice and form

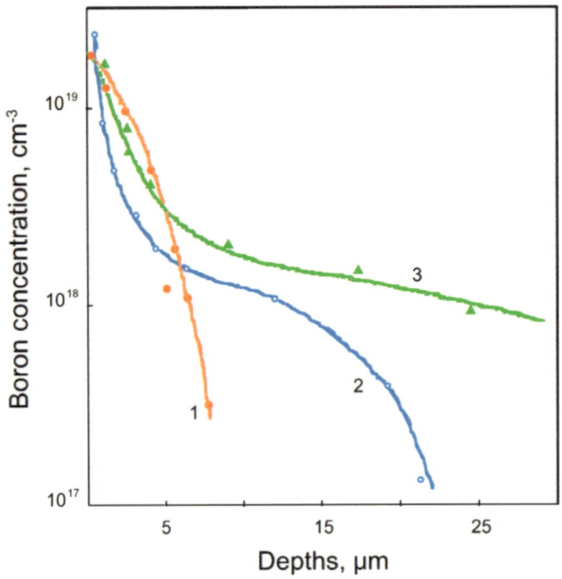

Figure 9.
Diffusion distributions of the isotope ^{10}B *in SiC: 1—constant-concentration diffusion in SiC layer* ($^{\Pi}$B) *with boron concentration* ($^{\Pi}$B) 5×10^{19} cm^3*; 2, 3—chemical diffusion in samples with different impurity concentration Al* (cm^{-3}}: 2×10^{16} (2); 5×10^{19} (3).

the inactive associates or clusters including several atoms of boron with mobile boron center. This assumption is proved by high binding energy of boron atoms as well as experimentally observed formation of second phase clusters enriched by boron impurity during diffusion annealing [42].

3.3.4 Boron diffusion from solid phase

Comparative study of boron diffusion from vapor and solid phase revealed essential dependence of boron diffusion parameters on the initial state of impurity [9, 61]. In our experiments in case of diffusion from solid phase, the source of impurity was usually located in preliminary grown epitaxial SiC layer doped by boron. In both cases surface concentration of boron was practically the same.

At short periods of annealing there were no observed considerable differences in the character of distribution and the rate of boron migration by diffusion from vapor and solid phase. However, with increase of the annealing period in case of solid phase diffusion, there was a distinct tendency of diffusion slowing down. A similar effect was observed in experiments on study redistribution of impurity by secondary annealing of diffusive samples.

It was found out that the rate of boron diffusion from solid phase strongly depends on growth conditions of doped epitaxial SiC layer, serving as a source of diffusing impurity. The increase of growth temperature of this layer by maintaining the constant concentration of boron in it leads to decrease of the diffusion rate. For example by diffusion of boron from SiC (B) layer grown at 2550°C, the coefficient of boron diffusion was 10^3–10^4 times lower than by diffusion from vapor.

Comparison of concentration profiles of diffusive samples before and after secondary annealing has shown that boron transition to crystal volume generally comes from heavily doped layer ($C_B > 10^{19}$ cm^{-3}), where a considerable part of boron impurity was in electrically inactive state.

The obtained results demonstrate that there are several boron centers in SiC with various diffusive mobility and relaxation time of processes connected with change of impurity centers state. This time is rather long than time of diffusive annealing. If the impurity center is situated in diffusion-active state (for example, is found in interstitial position or forms a mobile associate with a vacancy or interstitial atom), then the rate of diffusion will be raised. On the contrary if the impurity atoms are located in the lattice points and probability of forming of mobile centers is very low, then impurities diffusion rate sharply decreases. Obviously, the creation of the mobile centers of impurities by ion implantation can explain high diffusion rate of implanted atoms in comparison with vapor phase diffusion. The same reason explains acceleration of diffusion at oxidation and silicidation of the surface [62].

3.3.5 Mechanism of boron diffusion

The mechanism of boron diffusion is difficult to explain in the frames of interstitial model. Actually, the activation energy of boron diffusion is considerably high. Unlike Be we do not observe noticeable compensation of acceptor centers during boron diffusion in p-type SiC. It is difficult to explain the whole set of experimental data within a simple vacancy model. Within such mechanism it is difficult to understand the reason of the complicated character of diffusion distribution of boron; dependence of the diffusion coefficient on the surface concentration band; and the features of boron diffusion from solid phase. It is also necessary to take into account that boron in SiC mainly replaces silicon, and a simple mechanism of its diffusion along carbon vacancies is seemed improbable. Besides, the activation energy of boron diffusion in heavily doped p-SiC is much less than calculated [57].

Obviously, it should be assumed that a rapidly diffusing component is complex (B-Vc), in which concentration is limited by a flow of carbon vacancies injected from the heavily doped surface layer [24]. It is possible, provided that the concentration of generated vacancies is higher than equilibrium, defined, for example, by Shottky constant. It is important that the defect states injected in the crystal volume are significantly nonequilibrium. Such assumption is confirmed by strong dependence of boron diffusion coefficient on surface conditions and also by experimental evidence of nonstationarity of diffusion process from solid phase [61]. By increasing the density of dislocations which are vacancy traps the coefficient of boron diffusion decreases [63].

With increase of diffusion temperature one should expect relative reduction of contribution of this mechanism, both due to increase of concentration of thermal vacancies and probability of disintegration of rapidly diffusing associates. Model of boron diffusion was presented [64].

Another mechanism of boron diffusion in SiC was proposed in refs. [65, 66]. The authors offered that boron diffusion was realized by kick out mechanism, in which silicon interstitials play main role. Excess Si interstitials are created by ion implantation or by high boron impurity concentration [65].

3.4 Diffusion of aluminum and gallium

Aluminum and gallium are the shallowest acceptors in SiC. Therefore, interest to diffusion as a method of obtaining diode structures of various devices on the basis of SiC, such as power diodes, has been shown for a long time. Meanwhile, studying of these impurities diffusion meets great difficulties, mainly, because of their low diffusion mobility.

Our investigation of Al and Ga impurities diffusion was carried out from vapor phase in the range of temperatures of 1800–2400°C [67]. Metal Al or Ga was used as source of impurity. Diffusion distribution was studied by van der Pauw method. P-n junction method was also used. Then, the results [67] were supported by SIMS. The obtained results prove a rather complex mechanism of Al diffusion. Concentration distribution of Al has very abrupt near-surface region and a smoother volume region. The latter can be described by erfc function. The boundary concentration for the volume branch was 1×10^{18} cm^{-3}. It is not less than three orders lower than true surface concentration of Al.

For studying Al diffusion in a near-surface layer the Hall method [68] was used, and the measurements were taken by consecutive removal of layers 0.2–0.3 microns thick. As a result the diffusion coefficient for the near-surface branch has been defined as 5.6×10^{14}cm2/s at 2200°C. This value considerably exceeds the diffusion coefficient of Al found by the method of p-n junction in a sample with concentration of donors (nitrogen) Nd = 10^{19} cm^2/s.

The diffusion of Al from solid phase was also studied. The source of impurity was previously grown epitaxial SiC layer doped by Al. For study distribution of Al during diffusion annealing, the capacitive method was used. According to the obtained data the rate of Al migration from the epitaxial SiC(Al) layer was abnormally low (**Figure 7**). The diffusion coefficient for solid state diffusion was 10^3–10^4 times lower than by diffusion from vapor phase, and the activation energy (ΔE) was near 11 eV. Let us note for comparison that by diffusion from vapor, the activation energy of diffusion was $\Delta E = 6.1$ eV. These results correlate with data of van Opdorp [69], which show that diffusion coefficient from solid phase was 10^4 times lower than in case of vapor phase diffusion.

It is possible to conclude that diffusion of Al from vapor phase is carried out by migration of metastable (Al-Vc) complexes or deep Al centers. However, in the SiC(Al) crystals grown at high temperature, most of Al atoms, obviously, are situated in silicon units, and the concentration of rapidly diffusing (Al-V) associates is very low. In this case diffusion is possible only along silicon vacancies and demands considerable power expenses.

Diffusion profiles of Ga have no sharp near-surface region [24]. However, surface concentration of Ga (Cs = (3–5) $\times 10^{17}$ cm^{-3}) is much lower (30–50 times) than its contents in the epitaxial SiC layers grown under the same conditions. In general parameters of Ga and Al diffusion are close, which indicates identity of their diffusion mechanisms.

3.5 Diffusion of phosphorus

Phosphorus (P) is a donor impurity in silicon carbide. For estimation of diffusive mobility of P in SiC we used samples containing P entered by the method of transmutation doping on charged high-energy particles [70]. Change of the concentration profile of phosphorus in the sample by its high-temperature annealing was investigated. The results received by this technique reflect more adequately the process of solid-phase diffusion; as in this case the impurity centers are not entered through the phase boundary, and diffusion is carried out only due to thermal activation of atoms located in regular positions in the crystal volume. Thus, the probability of participation of impurity conditions generated on the surface in diffusive stream is minimized. The obtained data characterize diffusion in own semiconductor according to the inequality: $C_P \ll n_i$; (where C_P is concentration of the transmutated phosphorus; n_i—concentration of intrinsic charge carriers). By transmutation doping radiation defects are also entered. However, they are generally annealed at T < 2000°C, and their influence on the rate of impurity migration at higher temperatures can be neglected. 6H-SiC samples grown by the Lely method, mainly n-type of conductivity, doped by nitrogen were used. The concentration of noncompensated donors was (Nd-Na) = (2–4) $\times 10^{18}$ cm^{-3}. In the number of experiments the samples of p-SiC heavily doped by aluminum ($C_{Al} \ll 5 \times 10^{20}$ cm^{-3}) were also used.

For transmutation introduction of P SiC samples were irradiated by α-particles with the energy of 16 and 20 MeV at the current density j = 0.1 mcA/cm^2. For receiving sharper concentration profile of phosphorus, radiation was carried out at oblique incidence of the beam at the angle of 6–30°. Transmutation doping was carried out due to a nuclear reaction $^{29}Si(\alpha,p)^{32}P$. After that radiation samples were annealed in closed graphite containers in the atmosphere of argon at the temperature of 2000–2600°C. For evaporation reduction samples were located in an isothermal zone surrounded from all sides by fine SiC powder. The time of annealing varied from 30 min to 10 h, the thickness of the evaporated SiC layer in the course of annealing did not exceed 2–3 microns. The concentration profile of phosphorus was defined by measurement of residual β-activity of samples at consecutive removal of surface layers by the method of chemical etching in KOH alkali solution. The thickness of the removed layers was 1 micron. The profile of radioactive phosphorus was at the same time analyzed in a control sample that was not exposed to diffusive annealing. According to the received results, noticeable changes of the concentration profile were revealed only at temperatures of annealing above 2400°C. The diffusion coefficient was defined in the assumption that Fick's law was realized. For this purpose one-dimensional diffusion in a half-space with impenetrable border was considered.

The temperature dependence of P diffusion coefficient is given in **Figure 10**. Very high value of activation energy of phosphorus diffusion ΔE^P = 11.2 эВ attracts attention. It is higher than in case of self-diffusion of silicon (ΔE^{S1} = 8.18 эВ) and carbon (ΔE^C = − 8.2 эВ) in SiC [46]. Note that in the studied temperature range, D^P is closer to the self-diffusion coefficient of silicon. At the same time the coefficient of self-diffusion of carbon is 2–3 orders higher. Therefore, it is possible to assume that migration of phosphorus is carried out along carbon vacancies. However,

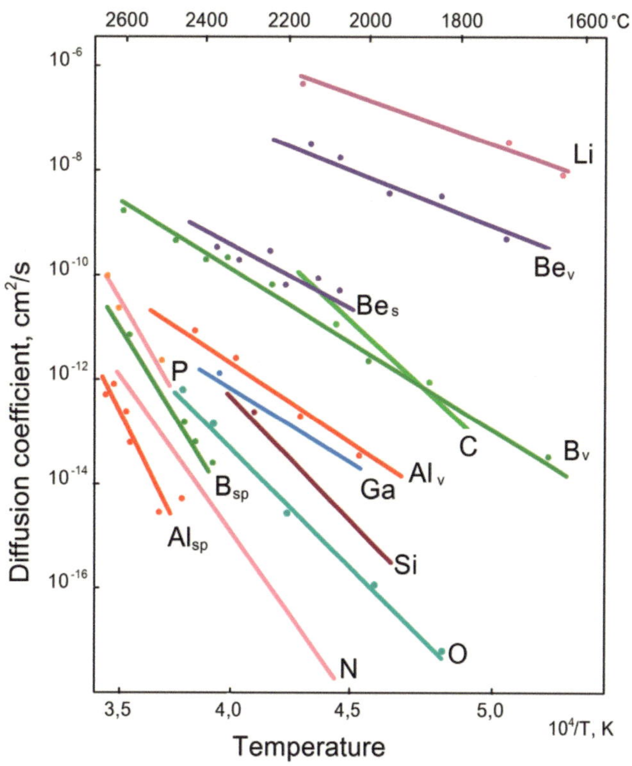

Figure 10.
Dependence of diffusion coefficients of impurities in SiC from vapor phase (Li, Be_v, Be_s, Bv, Al_v, Ga and N) and from SiC solid phase (P, B_{sv}, Al_{sv}). B_v, Al_v, Be_v—volume branch. Be_s—surface branch. Data on self-diffusion (C, Si) are taken from [46].

N	Impurity	D_0, cm^2/c	ΔE, eV	References
1	Nitrogen	$4.6–8.7 \times 10^4$	7.6–9.35	[71]
2	Phosphorous	1.3×10^{10}	11.6	[70]
3	Boron	3.2	5.1	[14]
4	Aluminum	8.0	6.1	[5]
5	Gallium	0.17	5.5	[19]
6	Beryllium Quick diffusion Slow diffusion	0.3 32	3.1 5.2	[3]
7	Lithium	1.2×10^{-3}	1.7	[66]

Table 3.
Parameters of impurity diffusion in SiC from vapor phase.

within the frames of such model, it is difficult to understand the reason of higher rate of phosphorus migration in highly doped p-SiC.

It is known that silicon vacancies in SiC are charged negatively; therefore, their concentration has to decrease with the increase of concentration of acceptor impurity, leading to reduction of phosphorus mobility. A reverse effect rather proving the mechanism of phosphorus diffusion with participation of carbon vacancies was experimentally observed. In this regard we will note that activation energy of phosphorus diffusion (ΔE) does not greatly differ from earlier defined ΔE value for nitrogen diffusion [71]. Nitrogen replaces carbon in SiC lattice and, obviously, migrates along carbon vacancies.

3.6 Diffusion of nitrogen and oxygen in SiC

Nitrogen and oxygen are donor impurities in SiC. Diffusion of nitrogen was carried out in p-type SiC samples doped by Al [11]. The temperature of diffusion varied from 1900 to 2500°C. Diffusion depth was found by Hall and p-n junction methods. Molecular nitrogen was used as diffusant. The temperature dependence of nitrogen diffusion coefficient is presented in **Figure 10**. It supports the early data of Kroko and Milnes [71].

The sources of oxygen impurity were CO_2 or SiO_2. Diffusion depth is found by measurement of thickness of the luminescence layer [72]. Higher oxygen diffusion can be explained by forming mobile complexes of impurity atoms with native defects. Unfortunately, the lack of reliable methods of identification of oxygen in SiC complicates the analysis of its diffusion mechanism.

The data of impurity diffusion in SiC are presented in **Table 3** and in **Figure 10**.

Let us note that solid state diffusion of B and Al impurities is characterized by slower rate than self-diffusion [46]. We consider that self-diffusion coefficients are actually lower than provided in the quoted works. Authors of the work [73] come to the same conclusion.

4. Conclusion

We have studied the processes of SiC doping by sublimation growth that allowed to receive the crystals and epitaxial SiC layers doped by various impurities and to define the condition of a number of impurity centers, including B [74, 75], Be [76], P [77], Sc [78], Mo [79], Er [80], Fe and Ni [81].

It is shown that the mechanism of SiC doping can be described within kinetic approach. High binding energy and, as a result, low concentration of equilibrium point defects cause low rate of solid-phase diffusion of the majority of impurity atoms. Noticeable diffusion at atom replacement from their regular positions occurs in SiC at the temperatures above 1800°C. Low diffusive mobility of the majority of impurity atoms leads to the fact that at actual growth rates, the external phase–crystal balance is not realized. It makes impossible consideration of the doping process within thermodynamic concepts. Doping of SiC by growth has a kinetic nature and reflects the external phase–crystal surface balance. Therefore, the content of impurities strongly depends on orientation of the growing surface. The instability of growth process leads to nonuniform doping. Uniform doping can be achieved by growth on a nonsingular surface. Small areas of octahedral and tetrahedral emptiness limit solubility of impurities with large atoms sizes.

By analysis of the doping processes it is necessary to consider high reactionary ability of native point defects, leading to formation of nonequilibrium stable associates with participation of both own defects and impurity atoms. These associates or clusters keep stability up to high temperatures. Formation of clusters is promoted by high mobility of own point defects. Interaction of impurity and matrix atoms promotes formation of precipitates. Such process can actively proceed on the surface of growing crystal that is facilitated by high surface diffusion. The interaction of impurity atoms with own defects leads to creation of a number of impurity states.

Acknowledgements

I thank A.A. Wolfson and O.P. Kazarova from the bottom of my heart for fruitful discussions and large help in manuscript preparation.

Author details

Evgeniy N. Mokhov
Ioffe Physical-Technical Institute of the Russian Academy of Sciences, St.Petersburg, Russia

*Address all correspondence to: mokhov@mail.ioffe.ru

References

[1] Suo H, Eto K, Ise T, Tokuda Y, Osawa H, Tsuchida H, et al. Crystal growth and evaluation of nitrogen and aluminum co-doped N-type 4H-SiC grown by physical vapor transport. Journal of Crystal Growth. 2018;**498**:224-229. DOI: 10.1016/2018.06.019

[2] Eto K, Suo H, Kato T, Okumura H. Growth of P-type 4H-SiC single crystals by physical vapor transport using aluminum and nitrogen co-doping. Journal of Crystal Growth. 2017;**470**:154-158. DOI: 10.1016/j.jcrysgro.2017.04.025

[3] Linnarsson MK, Janson MS, Zimmermann U, Svensson BG, Persson POA, Hultman L, et al. Solubility limits and precipitate formation in Al doped. Applied Physics Letters. 2001;**79**(13):2016-2018. DOI: 10.1063/1.1402160

[4] Vodakov YA, Lomakina GA, Mokhov EN. Non-stoichiometry and polytypism of silicon carbide. Soviet Solid State. 1982;**24**(5):780-784

[5] Onoue K, Nashikawa T, Katsumo M, Ohtani N, Yashizo H, Kanaya M. Nitrogen incorporation kinetics during the sublimation growth of 6H and 4H SiC. Japanese Journal of Applied Physics. 1996;**35**(1):2240-2243. DOI: 10.1143/JJAP.35.2240

[6] Linnarsson MK, Persson PO, Bleeichner H, Janson MS, Zummermann U, Andersson H, et al. Precipitate formation in heavy Al-doped 4H SiC layers. Materials Science Forum. 2001;**353-356**:583-586

[7] Ohtani N, Katsuno M, Takahasi J, Yashiro H. Impurity incorporation kinetics during modified Lely growth of SiC. Journal of Applied Physics. 1998;**83**(8):4487-4490. DOI: 10.1063/1.367234

[8] Vodakov YA, Mokhov EN. Point defects in silicon carbide. Institute of Physics Conference Series. 1994;**137**(3):197-206

[9] Mokhov EN. Impurities and native defects in silicon carbide in depending of conditions of growth, doping and relaxation annealing [Doctor Dissertation of doctor's of physical-mathematical sciences degree]. St. Petersburg, Rusia; 1998. pp. 583-586

[10] Vodakov YA, Roenkov AD, Ramm MG, Mokhov EN, YuN M. Use of Ta - container for sublimation growth and doping of SiC bulk crystals and epitaxial layers. Physica Status Solidi (b). 1997;**202**(1):177-200. DOI: 10.1002/1521-3951

[11] Vodakov YA, Mokhov EN. Diffusion and solubility of impurities in silicon carbide. In: Proceedings of the International Conference Silicon Carbide; 1973; South Carolina: South Carolina University Press; 1974. pp. 508-520

[12] Vodakov YA, Mokhov EN. Growth of semiconductor SiC. Patent: USSR N 403275 (1970); Fr: No 2 264 589 (1975); UK: No 1458445 (1977); FRG: No 24 09 005 (1977); USA: No 414572 (1979)

[13] Karpov SY, Makarov YN, Mokhov EN, Ramm MG, Ramm MS, Roenkov AD, et al. Analysis of SiC growth by sublimation sandwich – method. Journal of Crystal Growth. 1997;**173**:408-416. DOI: 10.1016/S0022-0248(96)00969-4

[14] Vodakov YA, Mokhov EN, Ramm MG, Roenkov AD. Epitaxial growth of SiC layers by sublimation sandwich method. Part 1. Growth kinetic. Kristall und Technik. 1979;**14**(6):729-740. DOI: 10.1002/crat.19790140618

[15] Mokhov EN, Wolfson AA. Growth of AlN and GaN crystals by sublimation.

Single Crystals of Electronic Materials: Growth and Properties. Duxford, United Kingdom: Woodhead Publishing; 2018. p. 401-456

[16] Avdeev OV, Chemekova TY, Mokhov EN, Nagalyuk SS, Helava H, Ramm MG, et al. Development of 2″ AlN substrates using SiC seeds. In: Modern Aspects of Bulk Crystal and Thin Film Preparation. United Kingdom: IntechOpen; 2012. pp. 213-262. DOI: 10.5772/29853

[17] Cavallotti C, Rossi F, Ravasio S, Masi MAA. Kinetic analysis of the growth and doping kinetics of the SiC chemical vapor deposition process. Industrial and Engineering Chemistry Research. 2014;**53**(22):9076-9087. DOI: 10.1021/ie403907w

[18] Mokhov EN, Ramm MG, Roenkov AD, Vodakov YA. SiC growth in Ta containers by the sublimation sandwich method. Journal of Crystal Growth. 1997;**181**:254-258. DOI: 10.1016/S0022-0248(97)00282-0

[19] Grillenberger J, Achtziger N, Sielemann R, Witthuhn W. Radiotracer identification of a Ta-related deep level in 4H–SiC. Journal of Applied Physics. 2000;**88**:3260. DOI: 10.1063/1.1289484

[20] Vodakov YA, Mokhov EN, Yuldashev GF, Usmanova MM. Neutron activation determination of solubility of impurities in SiC. Applied Nuclear Physics. 1973;(2):12-15

[21] Goncharov EE, Ryabova GG, Mokhov EE. Possibility of researching of semiconductor SiC by treck autogariography method. Isotopenpraxis. 1984;**12**:452-454

[22] Radovanova EI, Verenchikova RG, Vodakov YA. Determination of impurity concentration in n-type SiC by local breakdown of metal-semiconductor contact. Soviet Physics: Semiconductors. 1983;**17**(6):1115-1118

[23] Vodakov YA, Mokhov EN, Ramm MG, Roenkov AD. Doping peculiarities of SiC epitaxial layers grown by sublimation sandwich-method. In: Springer Proceedings in Physics. Washington, DC, USA: Materials of Conf. on Silicon Carbide; 1990. Vol. 56. 1992. pp. 329-334

[24] Vodakov YA, Lomakina GA, Mokhov EN, Radovanova EI, Sokolov VI, Usmanova MM, et al. Silicon carbide doped with gallium. Physica Status Solidi (A). 1976;**35**:37-43. DOI: 10.1002/pssa.2210350103

[25] Vodakov YA, Lomakina GA, Mokhov EN, Machmudov BS, Usmanova MM, Yuldashev GF. Silicon carbide, doped with Al and Ga. In book: Properties of Doping Semiconductors Moscow: Nauka. 1977;48-52

[26] Vodakov YA, Lomakina GA, Mokhov EN. Dependence of impurity solubility in SiC on crystallographic orientation and polytype. In: Doping of Semiconductors. Moscow: Nauka; 1982. pp. 230-233

[27] Mokhov EN, Usmanova MM, Yuldashev GF, Machmudov BS. Research of growth kinetics, solubility and polytypism of SiC epitaxial layers, doped phosphorus. Non-Organic Materials. 1981;**17**(2):258-261

[28] Vodakov YA, Mokhov EN, Roenkov AD, Saidbekov DT. Effect of crystallographic orientation on the polytype stabilization and transformation of the silicon carbide. Physica Status Solidi (A). 1979;**51**(1):209-215. DOI: 10.1002/pssa.2210510123

[29] EN DM, Maltsev AA. Doping 6H-SiC epitaxial layers, grown in excess of silicon with boron and nitrogen impurities. Soviet Technical Physics Letters. 1992;**18**(14):41-45

[30] Ramm MG, Mokhov EN, Verenchikova RG. Kinetic dependence of nitrogen impurity solubility in SiC. Non-Organic Materials. 1979;**15**(12):2233-2234

[31] Mokhov EN, Ramm MG, Vodakov YA. Growth of SiC epitaxial layers. Visokochictie Materiali. 1992;**3**:98-105

[32] Mokhov EN, Vodakov YA. Controlled growth of SiC and GaN by sublimation sandwich method. Institute of Physics Conference Series. 1997;**3**:177-182. No. 155

[33] Onoue K, Nashikawa T, Katsumo M, Ohtani N, Yashizo H, Kanaya M. Nitrogen incorporation kinetics during the sublimation growth of 6H and 4H SiC. Japanese Journal of Applied Physics. 1996;**35**(1):2240-2243

[34] Mokhov EN, Ramm MG, Roenkov AD, Fedorov VV, Verenchikova RG. Doping of the SiC epitaxial layers with nitrogen during sublimation sandwich metod in vacuum. Technical Physics Letters. 1990;**16**(14):33-37

[35] Mokhov EN, Verenchikova RG, Vodakov YA, Goncharov EE, Lomakina GA, Oding VG, et al. Research of specifics of doping during growth by sublimation sandwich - method. In: Doped Semiconductor Materials. Moskow: Nauka; 1985. pp. 45-52

[36] Mokhov EN, Usmanova MM, Yuldashev GF, Machmudov BS. Doping of SiC by IIIa group impurities. Non-Organic Materials. 1984;**20**(8):1383-1386

[37] Mokhov EN, Ramm MG, Roenkov AD, Vodakov YA, Verenchikova RG, Zabrodskii AG, et al. Capture of the impurities during SiC epitaxial growth from vapor phase. In book: Properties of Doped Semiconductor Materials. Moscow: Nauka. 1990;51-57

[38] Larkin DJ. SiC dopant incorporation control using site-competition CVD. Physica Status Solidi (b). 1997;**202**:305-320. DOI: 10.1002/1521-3951(199707)202

[39] Dorfman VD. Crystallography. 1968;**13**(1):140

[40] Mokhov EN, Didik VA, Kozlovski VV. The transfer of phosphorus impurity from sublimazed source in SiC epitaxial layer. Soviet Technical Physics Letters. 1992;**18**:659-662

[41] Vodakov YA, Jumaev N, Zverev BP, Lomakina GA, Mokhov EN, Oding VG, et al. Silicon carbide doped with boron. Soviet Physics: Semiconductors. 1977;**11**(2):214-217

[42] Mokhov EN, Zverev BP, Ramm MG, Usmanova MM. Surface boron distribution at diffusion in silicon carbide. Non-Organic Materials. 1980;**16**(12):2153-2156

[43] Yuldashev GF, Usmanova MM, Mokhov EN, Roenkov AD. Macro-segregation of doping impurities in SiC monocrystals and epitaxial layers. In book: Properties of Doped Semiconductor Materials. Moscow: Nauka. 1990;157-161

[44] Mokhov EN, Machmudov BS, Usmanova MM, Yuldashev GF. Solubility and macro-segregation of impurities in SiC. Soviet Technical Physics Letters. 1982;**8**(6):377-380

[45] Goncharov EE, Zubatov AG, Lomakina GA, Mokhov EN, Ryabova GG. Diffusion, solubility and EPR of Li in silicon carbide. Soviet Physics: Solid State. 1985;**27**(11):2398-2399

[46] Hon HH, Davis RF, Newburg DE. Journal of Materials Science. 1981;**16**(9):2485

[47] Linnarsson MK, Janson MS, Karlsson S, Shouer A, Nordell N,

Svensson BG. Diffusion of light elements in 4H and 6H SiC. Materials Science and Engineering: B. 1999;**61-62**:275-280. DOI: 10.1016/S0921-5107(98)00517-0

[48] Linnarsson MK, Doyle JP, Swenson BG. Diffusion of hydrogen in 6H SiC. MRS Proceedings. 1996;**423**:625-630. DOI: 10.1557/PROC-423-625

[49] Maslakovets YP, Mokhov EN, Vodakov YA, Lomakina GA. Diffusion Be in SiC. Soviet Physics: Solid State. 1968;**10**:634-638

[50] Vodakov YA, Mokhov EN, Oding VG. Interstitial diffusion of B and Be in SiC. Non-Organic Materials. 1983;**20**(7):1086-1088

[51] Vodakov YA, Lomakina GA, Mokhov EN, Oding VG, Radovanova EI. Beryllium as donor impurities in silicon carbide. Soviet Physics: Solid State. 1978;**20**(2):258-260

[52] Baranov PG, Mokhov EN. Electron paramagnetic resonance of deep boron in SiC. Institute of Physics Conference Series. 1996;**142**(1):293-296. DOI: 10.1088/0268-1242/11/4/005

[53] Bockstedte M, Mattausch A, Pankratov O. Different roles of carbon and silicon interstitials in the interstitial-mediated boron diffusion in SiC. Physics Review B. 2004;**70**:115203. DOI: 10.1103/PhysRevB.70.115203

[54] Vodakov YA, Mokhov EN, Reifman MB. Diffusion of boron and Al in n-SiC. Soviet Physics: Solid State. 1966;**8**:1040

[55] Mokhov EN, Vodakov YA, Lomakina GA, Oding VG, Kholuyanov GF, Semenov VV. Diffusion of boron in silicon carbide. Soviet Physics: Semiconductors. 1972;**6**(3):414-419

[56] Bolotnikov AV, Muzykov PG, Sudarshan TS. Investigation of two-branch boron diffusion from vapor phase in *n*-type 4H-SiC. Applied Physics Letters. 2008;**93**(5). DOI: 10.1063/1.2968306

[57] Mokhov EN, Goncharov EE, Ryabova GG. Boron diffusion in hole SiC. Soviet Physics Semiconductors. 1984;**18**(1):27-31

[58] Mokhov EN, Ramm MG, Usmanova MM, Zverev BP, Jumaev N. Influence of nitrogen on boron diffusion in silicon carbide. In: Problems of the Physics and Technology of Wide Bandgap Semiconductors. Leningrad: Nauka; 1979. pp. 316-320

[59] Vodakov YA, Lomakina GA, Mokhov EN, Oding VG. Effect of polytype structure on diffusion of impurities in SiC. Soviet Physics: Solid State. 1977;**19**(9):1812-1814

[60] Mokhov EN, Goncharov EE, Ryabova GG. Constant-concentration boron diffusion in silicon carbide. Soviet Physics: Solid State. 1988;**30**(1):140-142

[61] Goncharov EE, Mokhov EN, Ryabova GG. Specifics of boron diffusion in silicon carbide. In book: Properties of Doped Semiconductors. Moscow: Nauka. 1990;116-120

[62] Mendis S, Chin-Che T, Atabaev IG, Bakhtiyar G. The method of solid state impurity diffusion and doping in 4H-SiC. Journal of Fundamental Physical Sciences. 2013;**3**(4):75-78

[63] Vodakov YA, Mokhov EN, Prokhorov NA, Tredubova AS. Deceleration impurities diffusion by dislocations. Soviet Physics: Solid State. 1976;**16**(6):2101-2104

[64] Aleksandrov OV, Mokhov EN. Model boron diffusion in SiC from gas phase. Semiconductors. 2011;**46**(6):705-712. DOI: 10.1134/S1063782611060029

[65] Agarwal A et al. Boron-enhanced diffusion of boron: Physical mechanisms. Applied Physics Letters. 1999;**74**(16):2331-2333. DOI: 10.1063/1.123841

[66] Bracht H, Stolwijk NA, Laube M, Pensl G. Diffusion of boron in SiC: Evidence for the kick-out mechanism. Applied Physics Letters. 2000;**77**(20):3188-3190. DOI: 10.1063/1.1325390

[67] Mokhov EN, Vodakov YA, Lomakina GA. Diffusion of Al in silicon carbide. Soviet Physics: Solid State. 1969;**11**(2):415-416

[68] Lomakina GA, Vodakov YA, Mokhov EN, Oding VG, Kholuyanov GF. Comparative study of the electrical properties of 3 polytype of SiC. Physics of the Solid State. 1970;**12**(10):2356-2359

[69] Van Opdorp C. Abnormal diffusion of Al into SiC. Solid-State Electronics. 1971;**14**(7):813-825

[70] Mokhov E, Gornushkina ED, Didik VA, Kozlovski VV. Diffusion of phosphorus in silicon carbide. Physics of the Solid State. 1992;**34**(6):1043-1045

[71] Kroko LJ, Milnes AG. Diffusion of nitrogen in SiC doped Al. Solid-State Electronics. 1966;**9**(11-12):1125-1134

[72] Kholuyanov GF, Vodakov YA, Violin EE, Lomakina GA, Mokhov EN. Role of oxigen in blue and boron luminescence of silicon carbide. Soviet Physics: Semiconductors. 1971;**5**(1):32-36

[73] Rüschenschmidt K, Bracht H, Stolwijk NA, Laube M, Pensl G, Brandes GR. Self-diffusion in isotopically enriched silicon carbide and its correlation with dopant diffusion. Journal of Applied Physics. 2004;**96**(3):1458-1463. DOI:10.1063/1.1766101

[74] Duijn-Arnold A, Mol J, Verberk R, Schmidt J, Mokhov EN, Baranov PG. Spatial distribution of the electronic wave function of the shallow boron acceptor in 4H- and 6H-SiC. Physical Review B. 1999;**60**(23):15799-15809. DOI: 10.4028/www.scientific.net/MSF.338-342.799

[75] Van Duijn-Arnold A, Ikoma T, Poluektov OG, Baranov PG, Mokhov EN, Schmidt J. Electronic structure of the deep boron acceptor in boron-doped 6H- SiC. Physical Review B. 1998;**53**(3):1607-1619. DOI: 10.1103/PhysRevB.57.1607

[76] Van Duijn-Arnold A, Ikoma T, Poluektov OG, Baranov PG, Mokhov EN, Schmidt J. Electronic structure of the deep boron acceptor in boron-doped 6H- SiC. Physical Review B. 1998;**53**(3):1607-1619. DOI: 10.1103/PhysRevB.57.1607

[77] Baranov PG, Ber BY, Ilyin IV, Ionov AN, Mokhov EN, Muzafarova MV, et al. Peculiarities of neutron-transmutation phosphorous doping of 30-Si enriched Si and SiC crystals: Electron paramagnetic resonance study. Journal of Applied Physics. 2007;**102**:063713. DOI: 10.1063/1.2783884

[78] Spaeth J-M, Greulich-Weber S, März M, Mokhov EN, Kalabukhova EN. Electron paramagnetic resonance of the scandium acceptor in 4H and 6H silicon carbide. Physica B: Condensed Matter. 1999;**273-274**:667-671. DOI: 10.1088/0268-1242/15/1/310

[79] Dombrowski KF, Kunzer M, Kaufmann U, Schneider J, Baranov PG, Mokhov EN. Identification of molybdenum in 6H-SiC by magnetic resonance techniques. Physics Review B. 1996;**54**:7323-7327. DOI: 10.1103/PhysRevB.54.7323

[80] Baranov PG, Ilyn IV, Mokhov EN. Electron paramagnetic resonance of erbium in bulk SiC crystals. Solid State

Communications. 1997;**103**(5):291-295. DOI: 10.1016/S0038-1098(97)00161-0

[81] Baranov PG, Ilyin IV, Mokhov EN, Khramtsov VA. Identification of iron and nickel in 6H SiC crystal by electron paramagnetic resonance. Materials Science Forum. 2001;**353-356**:529-532. DOI: 10.1088/0268-1242/16/1/307. Trans. Tech. Publ. LTD, Switzerland Germany UK USA

Chapter 3

Numerical Analysis of Liquid Menisci in the EFG Technique

Sergei N. Rossolenko, Gleb M. Katyba, Irina N. Dolganova, Irina A. Shikunova, Dmitry O. Stryukov, Kirill I. Zaitsev and Vladimir N. Kurlov

Abstract

This chapter is devoted to the analysis of the behavior of the profile curves of the melt menisci for the sapphire crystal growth by edge-defined film-fed growth (EFG) technique. The menisci of the shaped crystals with capillary channels, fibers, and tubes (including cases of outer and inner circular menisci) are considered. Also, we investigated the profile curves of menisci both in the cases of the positive and negative angles between profile curve and the working edge of the die. The cases of outer and inner circular menisci of the tubular crystals and menisci at capillaries and fibers are considered.

Keywords: EFG technique, meniscus, profile curve, sapphire

1. Introduction

Single crystalline sapphire has high melting point, chemical inertness, impressive hardness, radiation and mechanical strength, high thermal shock resistance, and thermal conductivity. Sapphire also has high refractive index and a broad transmission band spanning the UV, visible, IR, THz, and microwave bands [1]. Such unique combination of physical and chemical properties makes sapphire an attractive material for various applications.

However, sapphire is difficult to shape because of its high hardness, which makes it difficult or impossible to obtain products of complex shape. In response to this problem, the edge-defined film-fed growth (EFG) technique [2] based on the Stepanov concept [3] was developed. This concept implies that the shape (or an element of the shape) to be produced is formed in the liquid state employing various effects, which enable the liquid to retain the shape. Then, the shape (or element of the shape) is converted to the solid state in appropriate crystallization condition. The method to form a melt column of a defined shape using a special die and to subsequently crystallize the melt column outside the walls of the vessel was suggested. The main idea of this concept is to limit the area of the liquid-free surface and its perturbations.

In the EFG technique, the crystal is grown from a melt film formed on the top of the melt-wettable die, which contains capillary channel (see **Figure 1**). The melt rises to the top of the die due to capillary forces, and the crystal growth proceeds at the top of the die. The edges of the die determine the shape of the meniscus and thus the cross section of the growing crystal. As a result, sapphire-shaped crystals

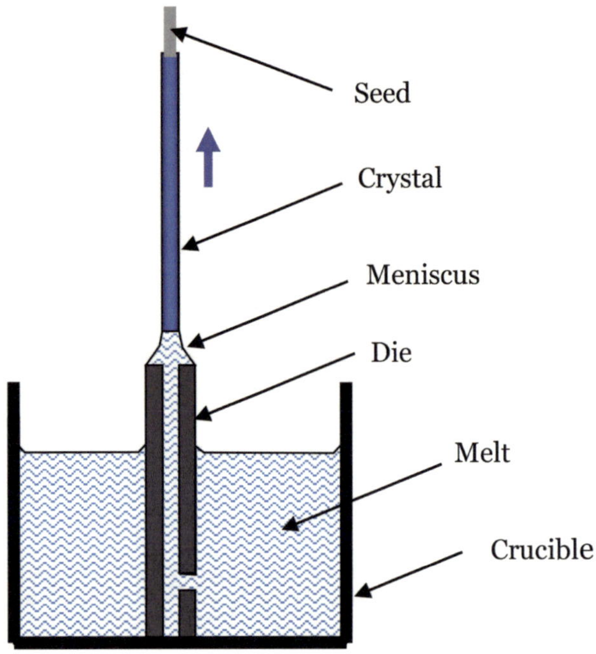

Figure 1.
Schematic of shaped crystal growth by the EFG technique.

of various constant and variable cross sections could be grown by the EFG technique due to its versatility for high-tech applications in optics, material science, biomedicine, etc. [4–10] with the relatively low production cost.

For the EFG technique, shape and quality of crystals are significantly defined by the form and position of the crystallization front and the shape of a liquid meniscus located between a crystal and a die. It is necessary to know the characteristics of the meniscus profile curves for their further use in dynamic models of the crystal-melt system, which are necessary for the development and optimization of automated systems for controlling the growth processes using a weight sensor.

Automated control systems make it possible to control not only the shape of the crystal but also its bulk and surface quality [11, 12], which is extremely important for expanding the fields of application of shaped crystals. In particular, preventing the formation of gas- and solid-phase inclusions in the volume of crystals and a significant improvement in the quality of the growth surface make it possible to use as-grown crystals in optics without additional surface treatment.

There are many publications devoted by the investigation of menisci shape evolution and their influence on the crystallization process. The approximate expression for the height of meniscus connected with the determined boundary conditions is represented in reference [13]. But this expression, usually being applied in Czochralski method, is difficult to use in EFG technique because of interconnected boundary conditions. Detailed consideration of this problem is represented in reference [14]. An approximate expression describing meniscus profile curve for the Czochralski method is given in reference [15]. Investigation of the influence of the negative outer pressure in the melt column of the meniscus on the limits of meniscus height in the crystal pulling of silicon by the EFG method is represented in reference [16]. Detailed research of the melt menisci, in the main, for the positive outer pressures, is given in reference [14]. The heights of menisci providing the implementation of the growth angle permanence condition and depending on the outer pressure for the sapphire-shaped crystal growth by the EFG method are given in reference [17]. In work [18] the equations for programmed masses of menisci (based on the integration of the Young-Laplace equation and approximate data from the weight sensor signal) for various forms of the crystals' cross section are described. These ones are used in the automated systems of the crystals' form and quality controlling [11] and in consideration of the dynamic models of the crystallization processes [19]. Non-cylindrical (almost quadratic) forms of the meniscus distinguishing for the growth of rare-earth molybdates by modified Czochralski and EFG methods [20] are investigated using numerical solution of the Young-Laplace equation submitted in Cartesian (non-cylindrical) system of coordinates [21]. Thermo-capillary numerical models also require solution of the equation [24–31]. In reference [32] the problem of mechanical stability of the liquid menisci is considered.

This chapter contains the results of study of the profile curves' formation, their properties, and dependence from the various external options and boundary conditions, which determine a meniscus form: outer pressure in the meniscus, height of the meniscus, contact angles between meniscus and working surface of the die, contact angles between meniscus crystal edge, and dimensions of the die and crystal. Variations of the ranges of external options and boundary conditions providing optimal conditions of the crystal growth process are discussed.

Figure 2 illustrates the scheme of meniscus zone for the tubular crystal growth. We consider one (right) side of the vertical cross section of the growing tubular crystal. There are two meniscus profile curves in this cross section—left (inner) and right (outer). The working surface of the die and the crystal edge contact with profile curve via the contact angles θ_d and θ_c, respectively. An angle between a profile line and a tangent line in the corresponding point defines the contact angle. We consider two types of the dies—with horizontal working surface (**Figure 2a**) and with sloped one (**Figure 2b**).

Estimation of the profile curves of the liquid menisci was made using the numerical analysis of the capillary Young-Laplace equation with various options and boundary conditions close to those of the EFG technique. The analysis of the melt-column shaping conditions was made for the case of catching meniscus at working edges of the wettable die. Special mention was paid for menisci providing stationary isotropic growth of sapphire crystal, i.e., profile curve should be satisfied with the condition of the growth angle permanency at the crystal edge (at the triple point). Specified boundary conditions for solving the Young-Laplace equation are close to the real ones resulting in the experimental processes of shaped crystal growth.

In this chapter the following results are presented:

- Analysis of the meniscus profile curves for various cases of catching with die, for various outer pressures, and various sizes of the crystal and the die
- Investigation of the shapes of menisci with negative and positive contact angles
- Analysis of the inner and outer menisci for the tubular crystal
- Analysis of the "planar" menisci for the case of the ribbon

It is well known [22] that the meniscus profile line $z(r)$ for cylindrical crystal and die is defined via capillary Young-Laplace equation in static approximation. For circular meniscus in dimensionless sight, it looks as follows:

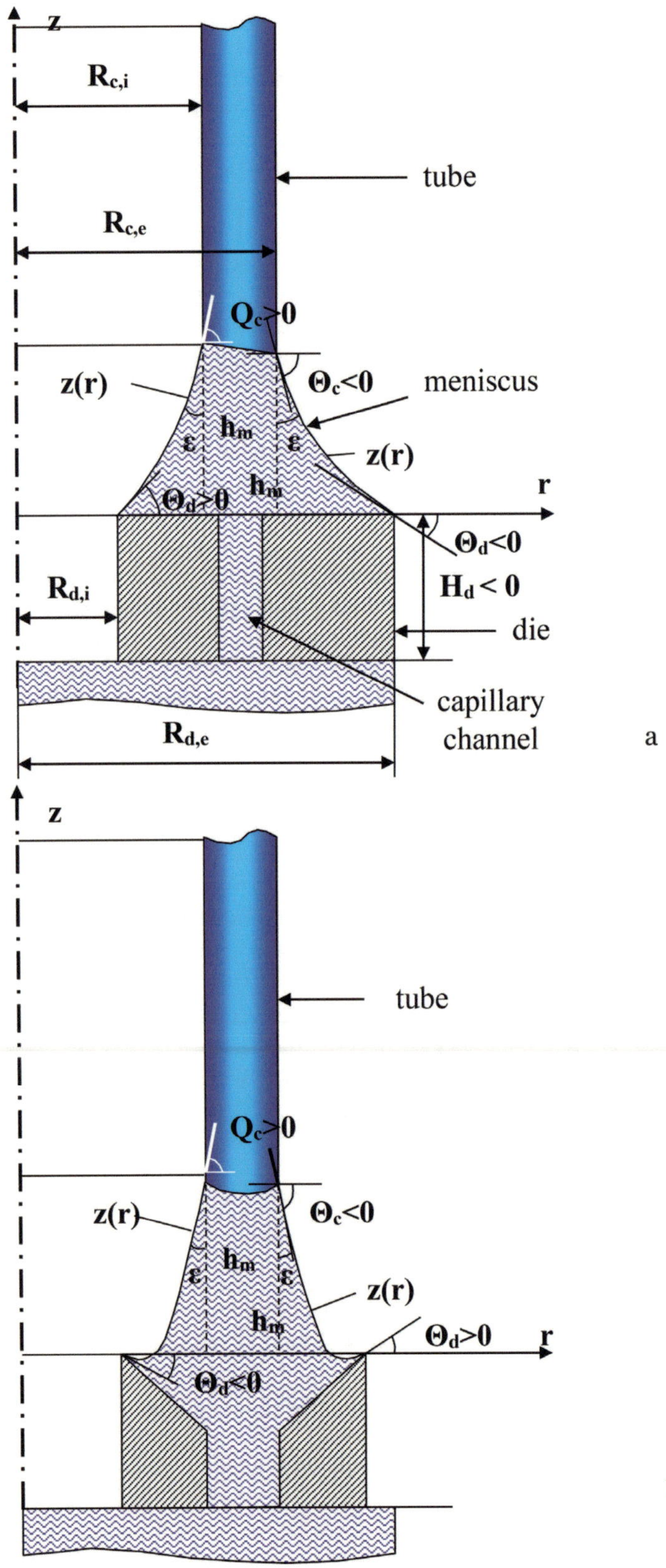

Figure 2.
Schematic of the crystal tube growth (a) with ordinary die and (b) with die having sloped working edges. The right vertical cross-section of the tube is shown.

$$z''r + z'\left(1 + z'^2\right) \pm 2(H_d - z)\left(1 + z'^2\right)^{3/2} r = 0 \tag{1}$$

where z is current meniscus height, r is current meniscus radius, and H_d is static outer pressure determined mainly by a difference between melt level in the crucible and meniscus base at the working surface of the die.

In sapphire crystal growth by the EFG technique, the base of the meniscus is usually above the melt surface in the crucible, i.e., $H_d < 0$. Thus, in the present chapter, we consider negative outer pressures in the meniscus. For solving the Young-Laplace equation in the case of normal stationary crystal growth, vertical coordinate $z(r)$ is predefined, and horizontal coordinate r is independent from z. Therefore, positive sign before the last part in Eq. (1) is considered. Taking into account the relation between z and r, we can analyze both negative and positive menisci contact angles θ_d between tangent meniscus profile line and horizontal line of the working surface of the die, i.e., the case of ambiguity along the vertical coordinate z is analyzed. In the case of the die with sloped working edge for the outer meniscus of the tubular crystal, a positive contact angle θ_d is possible (**Figure 2b**). In this case, the meniscus of the die is formed along this edge.

In the present chapter, menisci relating to the stationary crystal growth are considered and, hence, being formed without ambiguity along the r coordinate. Therefore, positive sign before the last part in Eq. (1) is considered.

Boundary conditions in our analysis for solving Eq. (1) can be determined as

$$z(R_d) = 0, -arctg\, z'\,(r_c) = \pi/2 - \varepsilon \tag{2}$$

where r_c is a radius of crystal, R_d is a radius of the working surface of the die, and ε is growth angle of the melt (for liquid $Al_2O_3 \approx 13°$ [17, 23]).

Zero height of the meniscus is predefined at the working edge of the die, and for the upper edge of the meniscus, the growth angle should keep permanent.

2. Outer circular menisci

2.1 Analysis of the menisci profile curves for various crystal radii

Numerical data shown in **Figures 3** and **4** were calculated for the outer pressure $H_d = -4$ in dimensionless units. The capillary constant of Al_2O_3 melt is approximately 6 mm [23]. Thus, this outer pressure corresponds to the difference of −24 mm between levels of the melt surface in the crucible and working edge of the die. It is close to data featured to real crystal growth process. Outer pressure is negative because of the upper location of the working edge of the die in comparison with the position of the melt surface in the crucible. The range of crystal radius changing was from 0.95 to 0.995 with step 0.005. We represent linear sizes in dimensionless units in this chapter. In this paragraph the radius of the working surface of the die was constant.

Figure 3a and **b** demonstrates profile curves of the menisci being calculated for the various values of the radius of the cylindrical crystal, i.e., outer right circular menisci of the crystal tube were considered.

As a result of the iterative process, the angle θ_d was automatically adjusted so that the angle θ_c of the meniscus at the point of contact with the crystal (at the triple point) satisfies the condition of constancy of the growth angle. Taking into account the growth angle of sapphire, the required slope angle of the meniscus to the horizontal line at the triple point should be equal to −77° with high accuracy.

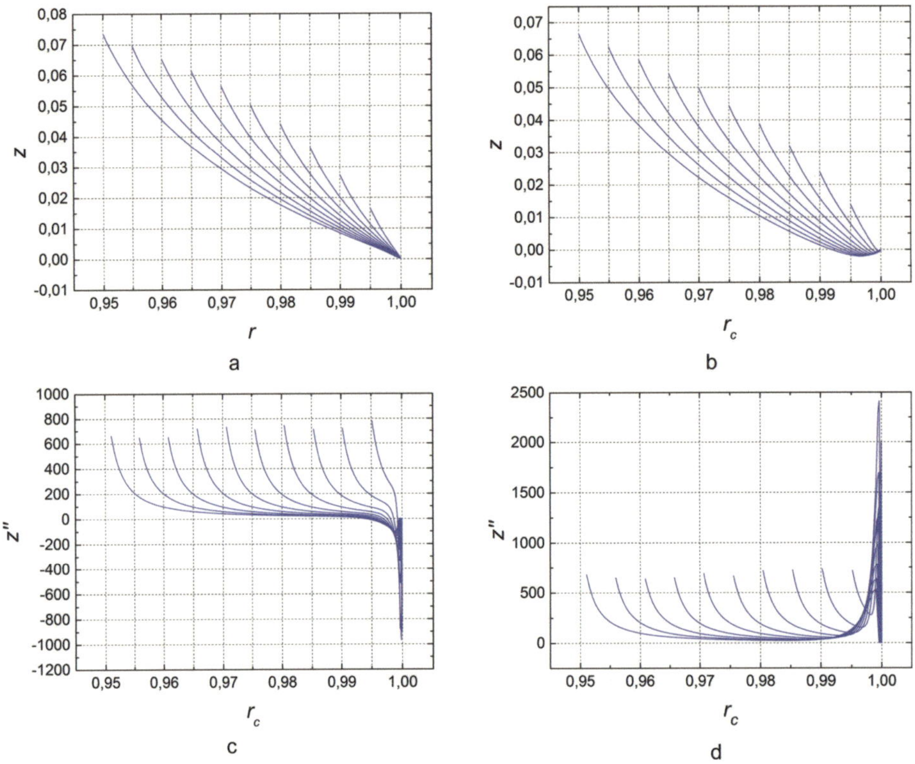

Figure 3.
(a, b) Profile curves of the outer menisci (a) for the negative angles θ_d and (b) for the positive angles θ_d and (c, d) second derivatives of the profile lines of the menisci (c) for the negative angles θ_d and (d) for the positive angles θ_d.

Profile curves depicted in **Figure 3a** correspond to the negative angles θ_d, while in **Figure 3b** to the positive angles θ_d. One can see different shapes of the menisci profile curves in these two cases.

We observe the decrease of the meniscus height with crystal radius increase (**Figure 3a**). With the oncoming of the crystal to the edge of the die, meniscus curves move upward from the die more intensively. There is an upward convexity of the meniscus curves near the die. After the inflection point, a convexity becomes downward.

Meniscus profile curves with positive angles θ_d for the die with sloped working edges are shown in **Figure 3b**. **Figure 3b** demonstrates that the points on the profile curve are displaced down in close proximity to the die and then, as a rule, are displaced upward, toward the edge of the crystal. In both cases, when the radius of the crystal increases, the behavior of the menisci is similar. Thus, if we change the angle θ_d from $-\pi/2$ to $+\pi/2$, the Young-Laplace equation has two solutions satisfying the boundary conditions (2): for the positive and negative angles θ_d.

The second derivatives of the meniscus profile curves are demonstrated in **Figure 3c** and **d**. The second derivatives characterize a curve convexity direction. Negative second derivatives near the die for negative contact angle (**Figure 3c**) correspond to a bulge upward. With the rise of the meniscus height, the second derivative becomes positive. This corresponds to a bulge of the meniscus profile curve downward. Thus, the second derivatives grow as the profile curve approaches the edge of the crystal. This shows that the convexity of the profile curves increases near the edge of the crystal.

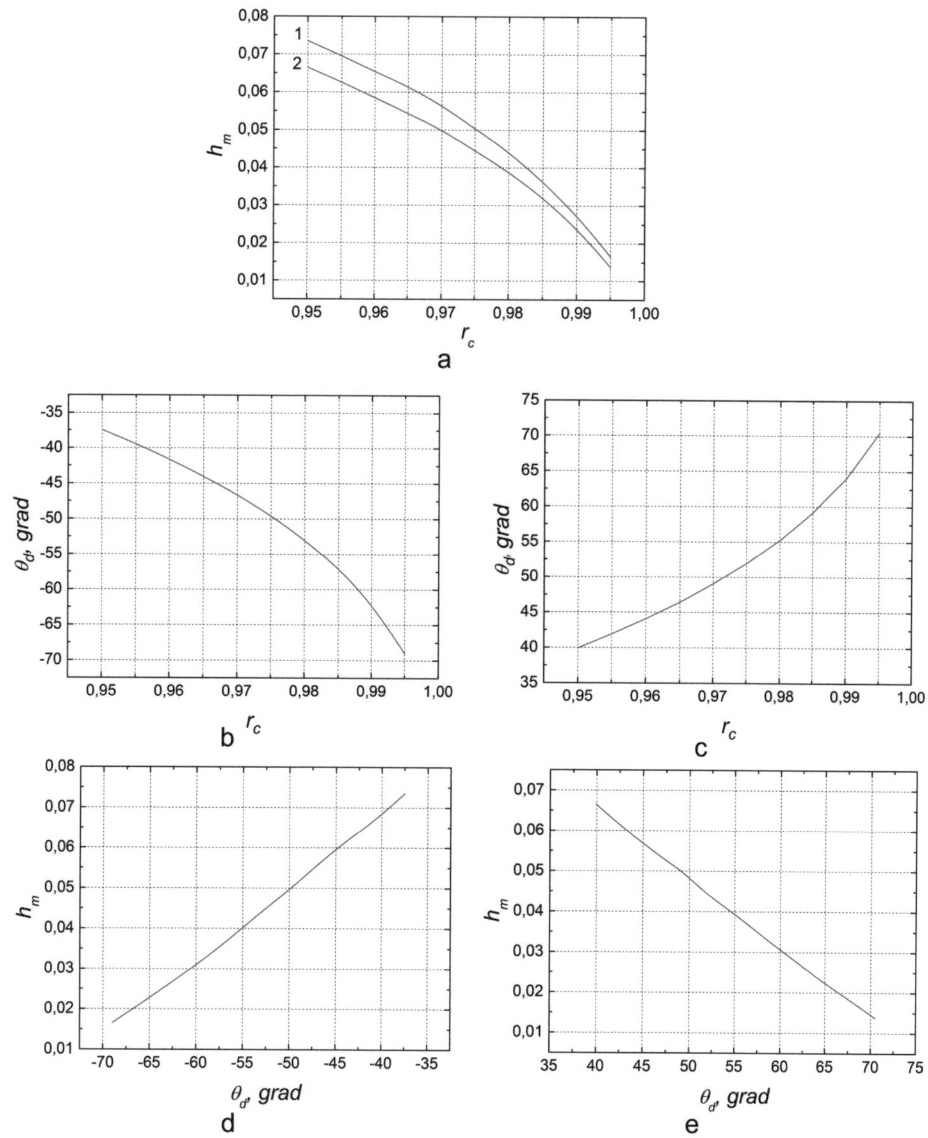

Figure 4.
(a) The curves of the meniscus height depending on the crystal radius (first line) for the contact angles $\theta_d < 0$ and (second line) for the contact angles $\theta_d > 0$, (b, c) the curves of the contact angle θ_d depending on the crystal radius (b) for the angles $\theta_d < 0$ and (c) for the angles $\theta_d > 0$, (d, e) the lines of meniscus height depending on the contact angle θ_d (d) for the angles $\theta_d < 0$ and (e) for the angles $\theta_d > 0$.

The second derivatives of the meniscus profile curves for positive angles θ_d (see **Figure 3d**) have a positive sign everywhere along the current radii of the meniscus. The values of the second derivatives near the crystal are significantly larger than at a sufficient distance from it. In the case of positive angles θ_d, the inflection point is absent, and menisci have convexity everywhere in melt direction.

Figure 4a demonstrates dependence of the height of the meniscus on the radii of the crystal. We observe diminishing meniscus height with crystal radius rise (**Figure 4a**) and its oncoming to the edge of the die. Decrease of the meniscus height corresponds to smaller height of the interface boundary. This corresponds to colder thermal zone of the system "melt-crystal." In a growth process, it is necessary to choose some middle position of the interface boundary corresponding to the optimal conditions for crystal formation.

Figure 5 shows the surface of dependence of the outer meniscus height on the outer pressure and the radius of the die for the case of the negative angle θ_d and for the crystal tube outer radius 0.97 (in capillary constants).

2.2 Influence of various outer pressures on the menisci profile curves

We have made the calculations for various pressures being changed from −8 to 0 with step 1 in capillary constants (dimensionless units) via the constant die and crystal radii equal to 1 and 0.97, respectively. The requirement of growth angle permanence was satisfied during calculations.

With the rise of outer pressure in the range from $H_d = -8$ to $H_d = 0$ the meniscus height increases (**Figure 6a, b**). It takes place because of diminishing the force pressing meniscus down to the die.

As shown in **Figure 7a**, meniscus height decreases with the rise of absolute value of outer pressure for negative contact angles as well as for positive ones. Therefore, **Figure 7** shows the features of the meniscus profile lines only for negative boundary angles θ_d.

Diminishing the absolute value of the outer pressure increases the absolute value of the contact angle θ_d (**Figure 7b**). To obtain the appropriate (higher) meniscus height, the larger absolute values of contact angles θ_d are required (**Figure 7a** and **b**). The function of the meniscus height depending on the boundary angle θ_d is almost linear (**Figure 7c**).

2.3 Influence of various boundary angles θ_d on the menisci profile curves

We have made analysis for the various boundary angles θ_d, while the die and crystal radii were constant. The die radius was equal to 1, crystal radius was equal

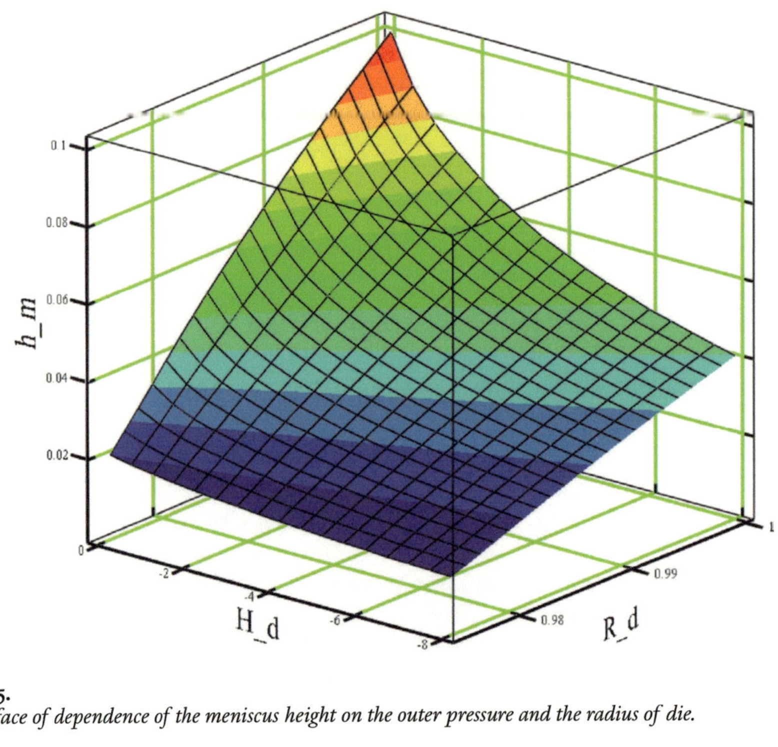

Figure 5.
The surface of dependence of the meniscus height on the outer pressure and the radius of die.

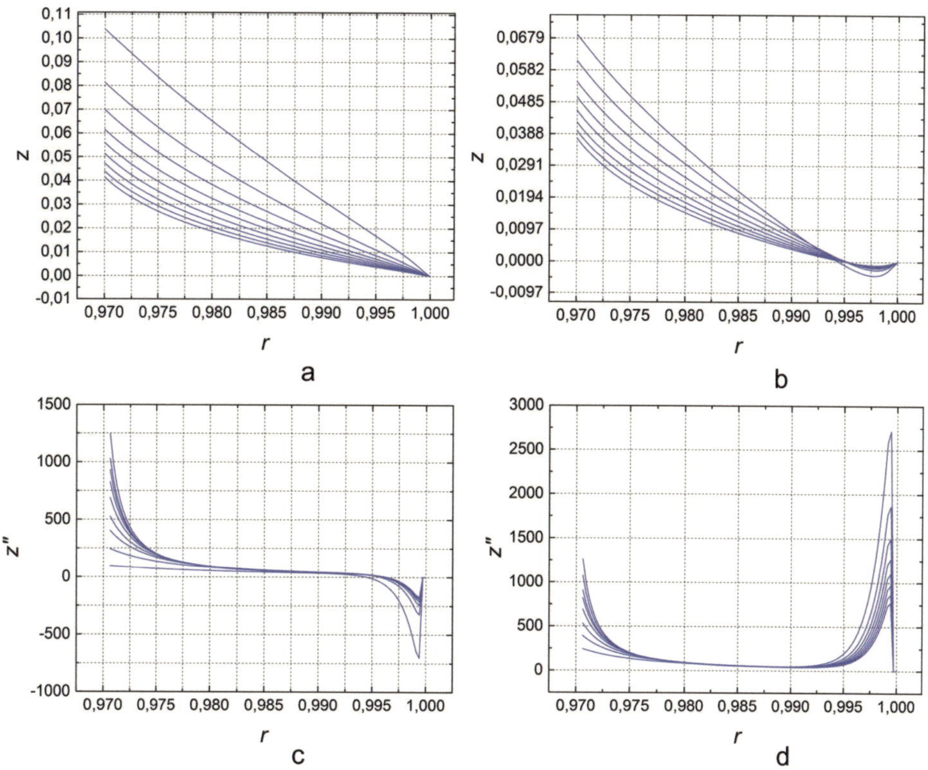

Figure 6.
(a, b) Profile curves of the menisci for the various outer pressures (a) for the negative angles θ_d and (b) for the positive angles θ_d and (c, d) second derivatives of the profile curves of the menisci for the various outer pressures (c) for the negative angles θ_d and (d) for the positive angles θ_d.

to 0.97, and outer pressure was equal to −4, in dimensionless units (capillary constants). Requirement of the growth angle permanence at triple point was not satisfied. We have used the next boundary condition:

$$z(R_d) = 0, arctg\, z'\,(R_d) = \theta_d \tag{3}$$

Figure 8a shows that the large absolute values of the boundary angles θ_d result in significantly higher menisci. Diminishing and rise of line segments (**Figure 8a**) correspond to the case above. Derivatives of the line decrease and rise is significant. Hence, the range of the menisci heights being sufficient for the stable growth and being satisfied with the condition of the growth angle permanence is sufficiently narrow. **Figure 8b** demonstrates that the ranges of boundary angles being almost real are approximately from −50 to −30° for the negative boundary angles θ_d and from +30 to +50° for the positive ones.

Under relatively small absolute values of the boundary angles (<30°), θ_d menisci heights are very small and unreal in a course of the sapphire crystal pulling by EFG technique.

2.4 Influence of various values of die and crystal radii (with constant difference between them) on the features of menisci

We have considered simultaneous changing of the die and crystal radii with constant distance between them, which was equal to 0.03 in dimensionless units (capillary constants).

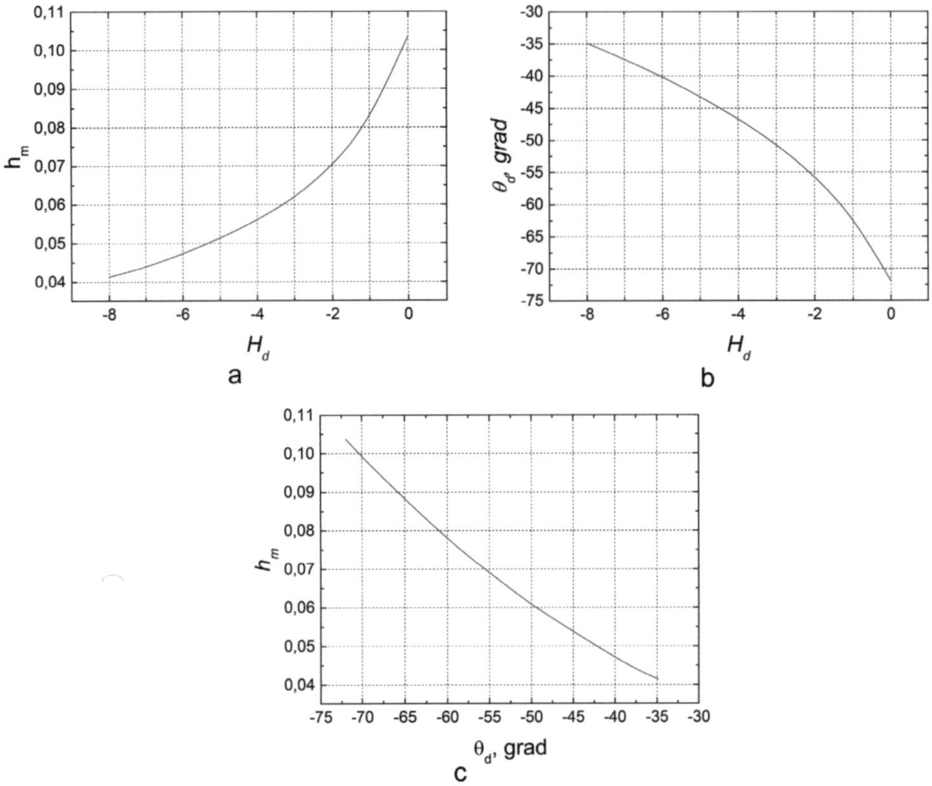

Figure 7.
(a) Function of the meniscus height depending on the outer pressure, (b) function of the boundary angle θ_d depending on the outer pressure, and (c) function of the meniscus height depending on the boundary angle θ_d.

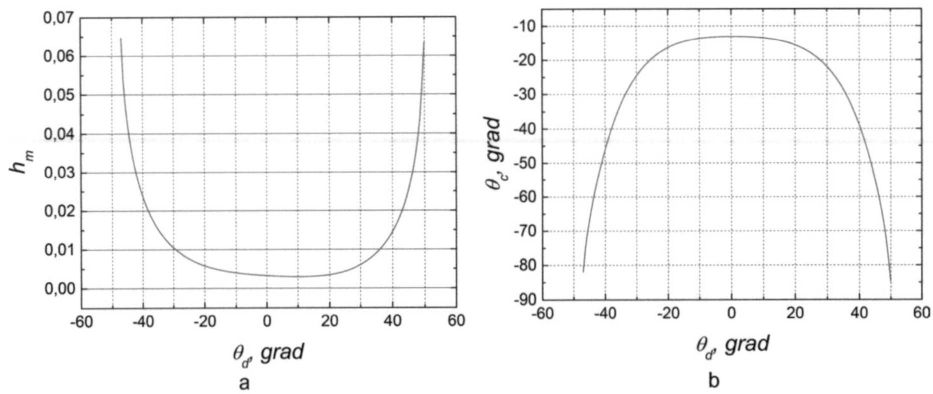

Figure 8.
(a) Function of the meniscus height depending on the boundary angle θ_d and (b) function of the boundary angle θ_c depending on the boundary angle θ_d.

The die radius was changing from 0.1 to 20, and the crystal radius was changing from 0.07 to 19.97. For sapphire it corresponds to the change from 0.42 mm to approximately 120 mm of crystal radius. We took into account the requirement of the growth angle permanence at the crystal edge.

Figure 9a shows that the strong changing of the meniscus height takes place at the die and crystal radii change at sizes of the capillary constant. The same situation is also for the change of θ_d (**Figure 9b**).

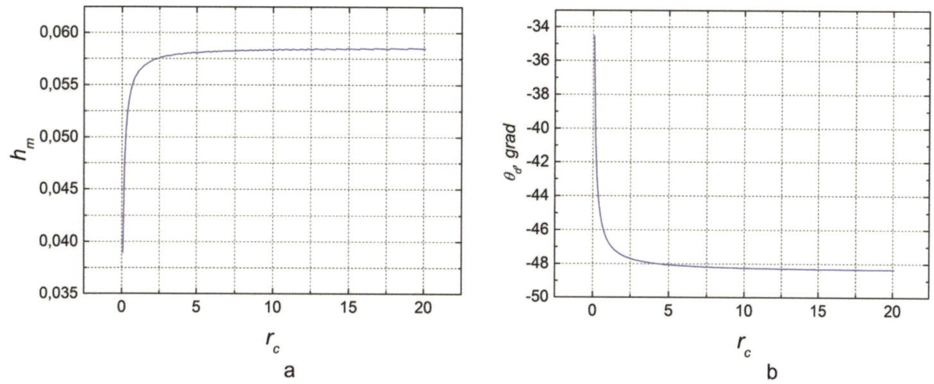

Figure 9.
(a) Function of the meniscus height depending on the crystal (and die) radius for the tubular crystal and (b) function of the contact angle θ_d depending on the crystal (and die) radius for the tubular crystal.

3. Inner menisci for the tubular crystal

We have analyzed the inner menisci for the tubular crystal with similar data as for outer menisci. **Figure 10** shows typical profile lines of the inner menisci. There are some differences connected with the fact that azimuthal curvature (second part of the capillary Eq. (1)) has positive sign for the inner meniscus and negative sign in the case of outer meniscus.

This curvature enlarges the current meniscus height due to the positive sign of the azimuthal curvature for the inner menisci. Profile curves of the inner menisci are a little bit higher than outer menisci due to the positive sign of the azimuthal curvature. This results in the presence of additional compressing factor in the inner menisci.

We have also considered simultaneous change of the die and crystal radii with constant distance between them. Outer pressure H_d was equal to −4 in dimensionless units (capillary constants). Requirement of the growth angle permanence in triple point was implemented.

The dependence of the meniscus height on the crystal and die radii (**Figure 11a**) has difference in comparison with the one in the case of the outer circular menisci and has opposite behavior. The line of the inner menisci height is slightly diminishing, but outer menisci height rises. This can be explained by the opposite signs of the azimuthal curvature.

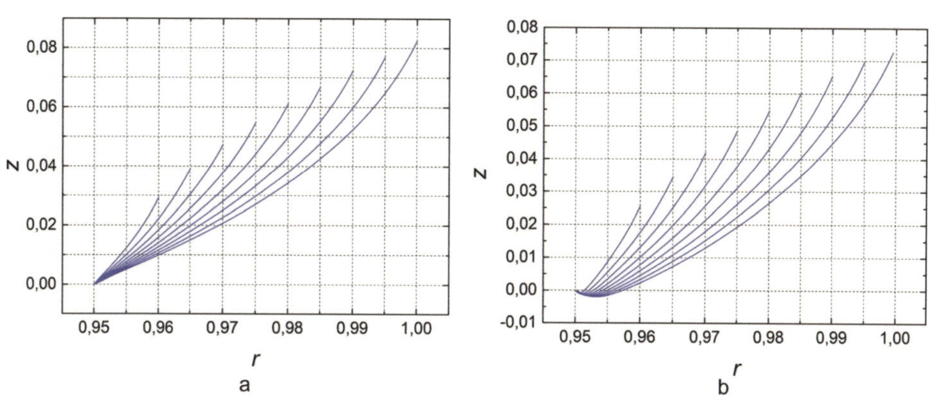

Figure 10.
(a, b) Profile lines of the inner menisci (a) for the positive angles θ_d and (b) for the negative angles θ_d.

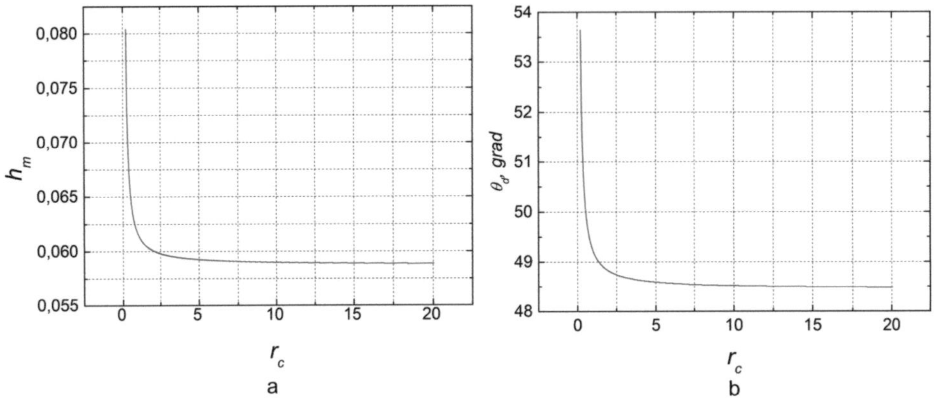

Figure 11.
(a) Function of the meniscus height depending on the crystal (and die) radius for the inner circular menisci and (b) function of the contact angle θ_d depending on the crystal (and die) radius for the inner circular menisci.

We have considered also "planar" menisci being formed at the wide side of the crystal ribbon. The Young-Laplace equation was analyzed with zero azimuthal curvature. Using the same modeling options (as for circular menisci), we have found that the heights of "planar" menisci are a little bit higher than the heights for the outer circular menisci and a little bit lower than the heights for the inner menisci. This can be explained by zero azimuthal curvature. The influence of the various outer pressures, various crystal ribbon width, and other factors is the same as for the circular menisci.

4. Application of menisci analysis to automated control in EFG technique

The programmed expression of observing for weight sensor is necessary for automated control in EFG technique. At periods of control, the mass being calculated via this expression is compared with real signal of the weight sensor. Obtained deviation is used for calculation of regulating impact in the feedback closed loop of control process.

For the stationary tubular crystal growth with quasi-planar crystallization front, the meniscus part of the observing expression should be written as follows [12, 18]:

$$\begin{aligned} M_m = \pi\rho_L\left(r_{c,e}^2 - r_{c,i}^2\right)\left(h_{m,e} + h_{m,i}\right)/2 - 2\pi\rho_L a^2 r_{c,e} sin\theta_{c,e} + 2\pi\rho_L a^2 r_{c,i} sin\theta_{c,i} \\ + 2\pi\rho_L a^2 R_{d,e} sin\theta_{d,e} - 2\pi\rho_L a^2 R_{d,i} sin\theta_{d,i} - 2\pi\rho_L\left(R_{d,e}^2 - R_{d,i}^2\right) H_d \end{aligned} \quad (4)$$

Here, $r_{c,i}$ is a radius of the tube inner side, $r_{c,e}$ is a radius of the tube outer side, $R_{d,i}$ is a radius of the die inner working edge, $R_{d,e}$ is a radius of the die outer working edge, $h_{m,i}$ is an inner meniscus height, $h_{m,e}$ is an outer meniscus height, $\theta_{c,i}$ is a contact angle of the inner meniscus with the crystal edge, $\theta_{c,e}$ is a contact angle of the outer meniscus with the crystal edge, $\theta_{d,i}$ is a contact angle of the inner meniscus with die working surface, $\theta_{d,e}$ is a contact angle of the outer meniscus with the die working surface, ρ_L is a density of the melt, a is a capillary constant, and H_d is an outer static pressure.

The angle θ_c is connected with growth angle and specific for each liquid material. The outer pressure can be calculated from the dimensions of crucible and die and

Figure 12.
Sapphire tubular crystals of 55 mm in outer diameter grown by EFG technique using automated control system.

total melt charge in the crucible. The programmed menisci heights and angles of contact with working surface of die can be calculated from solution of the Young-Laplace equation.

For the known outer pressure and growth angle, we can find menisci heights and angles of contact with the die, using **Figures 7b** and **8a**. Similar estimations can be made for inner as well as for outer menisci.

The optimal heights of inner and outer menisci and, hence, optimal conditions at the crystallization front are necessary to grow sapphire tubular crystals of high quality. To reach this purpose, die top design and special thermal shields are used [33].

We have grown large-scale sapphire tubular crystals of 55 mm in the outer diameter (**Figure 12**) using optimal values of menisci heights and contact angles in the automated control system.

5. Small inner menisci for tubular crystals (at capillary)

We have analyzed small inner menisci of capillaries for various crystal radii for the outer pressure $H_d = -4$ in dimensionless units (capillary constants). We changed the crystal radius from 0.05 to 0.0486 with step −0.0002. For sapphire it corresponds to the change from 0.3 to 0.2916 mm with step −0.0012 mm. The radius of the die edge was equal to 0.0485 (in dimensionless units) or 0.291 mm. We have made calculations for the positive contact angles θ_d.

As shown in **Figure 13a**, menisci profile lines have another character in comparison with lines demonstrated in **Figure 3a** shown above for sufficiently "large" menisci. As mentioned above, it is connected with the fact that azimuthal curvature has positive sign in the case of inner meniscus and negative sign in the case of outer meniscus.

As shown in **Figure 13a**, capillary menisci have an upward convexity. It takes place because of the very small weight and sufficiently large value of azimuthal convexity. The heights of capillary menisci are sufficiently small because of the small sizes of crystal radii (**Figure 13b**).

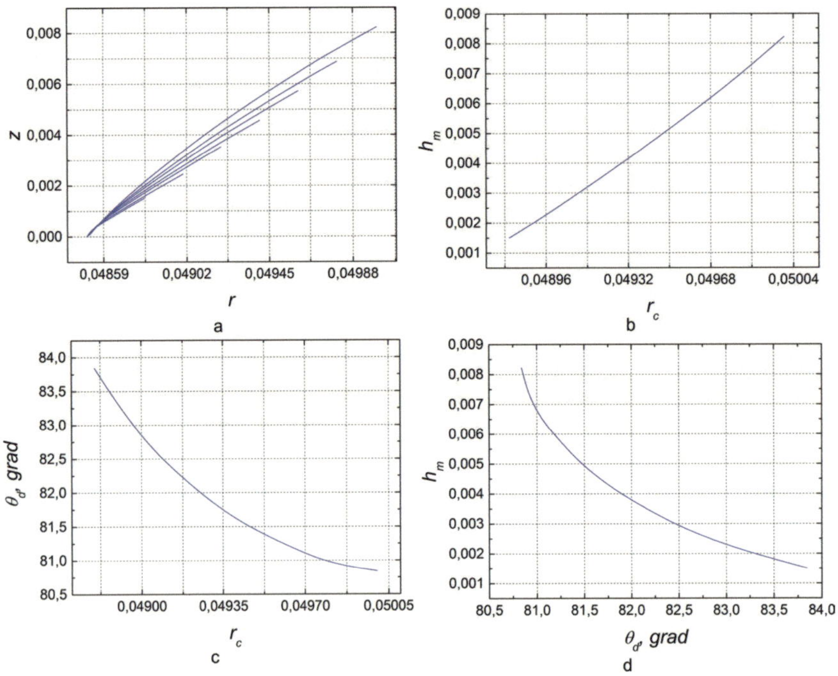

Figure 13.
(a) Profile lines of the small inner capillary menisci at the various crystal radii, (b) function of height of the capillary menisci depending on capillary radius, (c) function of the angle θ_d depending on the capillary radius, and (d) function of height of the capillary menisci depending on the angle θ_d. Positive angles θ_d are considered.

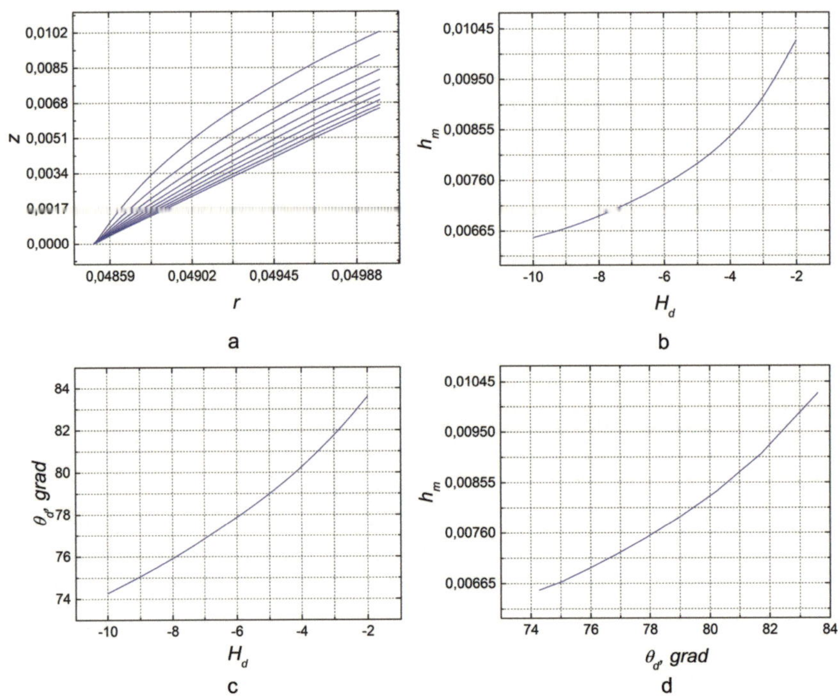

Figure 14.
(a) Profile lines of the small inner menisci (at capillaries) for the various outer pressures, (b) function of the meniscus height depending on the outer pressure, (c) function of the angle θ_d depending on the outer pressure, and (d) function of the meniscus height depending on the angle θ_d at various outer pressures.

For capillary menisci, absolute values of the contact angles (**Figure 13c**) are significantly larger than for the "large" menisci. For the small menisci, larger angles θ_d are required, because the form of small profile lines has convexity directed upward, differing by this from the form of the "large" menisci.

For the small menisci, dependence of the meniscus height on the angle θ_d (**Figure 13d**) is nonlinear compared with "large" menisci. It takes place due to the upward convexity of the capillary menisci.

The die radius was 0.0485, and crystal radius was 0.05 in investigation of the capillary profile lines for the various outer pressures. The value of the outer pressure was changed from −10 to −2 with step 1. Calculations have been fulfilled for the positive angles θ_d.

As demonstrated in **Figure 14**, the capillary meniscus height becomes smaller with rise of the absolute value of the outer pressure as in the case of "large" menisci.

6. Small menisci for fibers

We have made the investigation of the profile lines of the small (outer) fiber menisci variating the crystal radii under the outer pressure $H_d = -4$ (in dimensionless units). We were changing the crystal radius from 0.0483 to 0.0499 with step 0.0002. For sapphire it corresponds to the change from 0.290 to 0.2994 mm with step 0.0012 mm. The radius of the die working surface is equal to 0.05 (in dimensionless units) or approximately 0.3 mm. The negative boundary angles θ_d have been considered.

Due to another sign of the azimuthal curvature, the form of the fiber menisci strongly differs from the capillary ones (**Figure 15**).

We have made analogous numerical analysis for small fiber menisci as for the case of the "large" outer menisci (**Figure 5**). As a whole, behavior of the small fiber menisci and their features are similar to behavior of the "large" outer menisci for the tubular crystals (**Figure 16**).

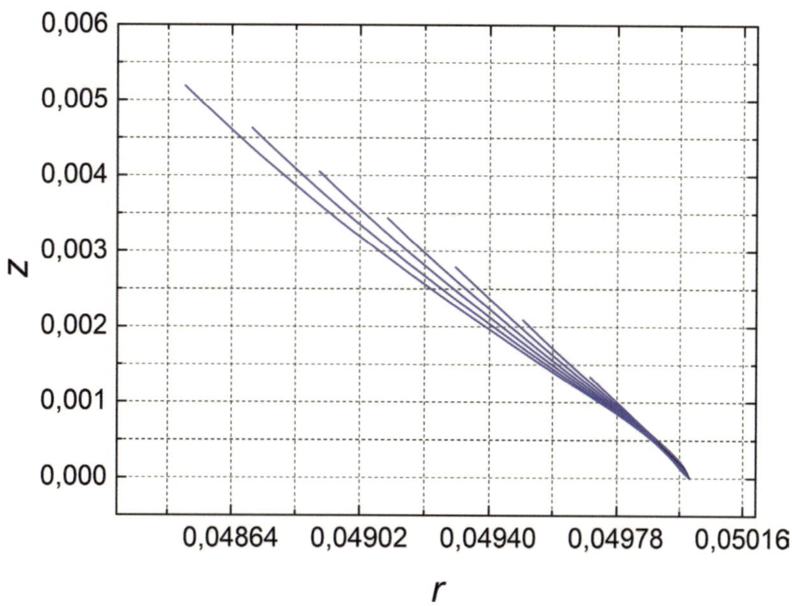

Figure 15.
Profile lines of the small (outer) fiber menisci at various crystal fiber radii. Negative angles θ_d are considered.

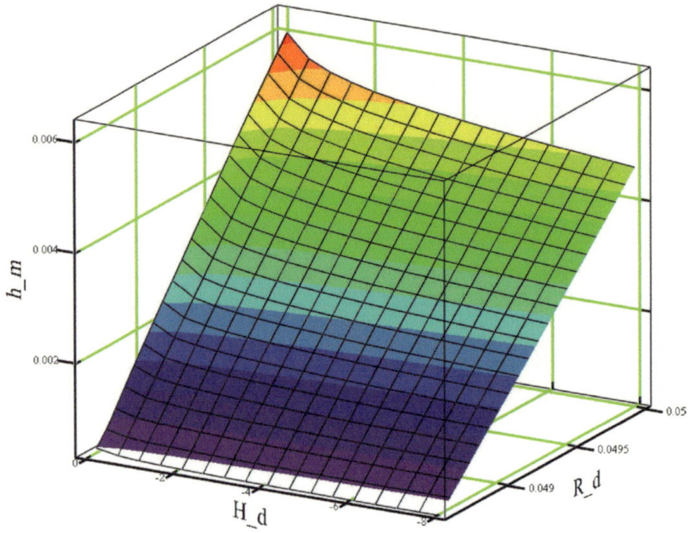

Figure 16.
The surface of dependence of the small meniscus height on the outer pressure and the radius of die.

7. Application of small menisci analysis to sapphire fibers and capillaries' growth using automated control

The data calculated from numerical solution of the Young-Laplace equation for the fibers and capillaries as well as for the large crystals are used in the programmed expression of observation of the weight signal. Due to the deviation between the programmed data and the weight signal, there are first and second derivatives of the deviation calculated that form the signal for heating power by the proportional-integral-differential (PID) procedure.

Automated system also observes the amplitude of the first derivative of the weight signal deviation. The sufficiently strong amplitude corresponds to over-cooling in the thermal zone, and sufficiently small amplitude corresponds to its overheating [12]. Thus, this amplitude should have optimal range of changing. For fibers and capillaries (inclusive the multi-crystal fibers pulling), this range is

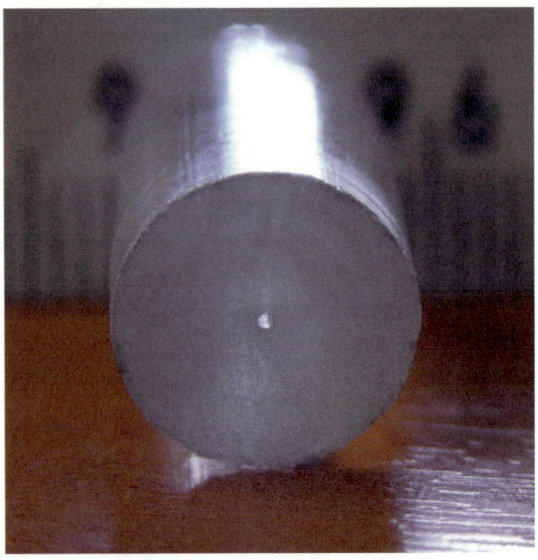

Figure 17.
Sapphire rod of 10 mm in outer diameter with capillary channel of 450 μm.

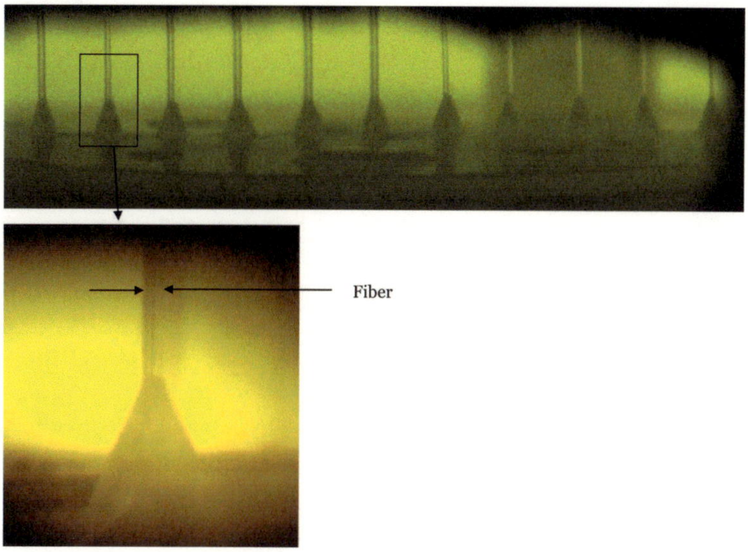

Figure 18.
Sapphire fibers of 150–300 μm in diameter grown by multi-crystal growth process.

sufficiently narrow. The need of maintaining the narrow range of these oscillations' amplitude requires the use of the second control loop tuning the first PID loop by the appropriate change of the programmed radius of the crystal or middle radii of the crystals in multi-crystal pulling.

An example of the automated growth of the capillary channel in the body of the bulk crystal is demonstrated in **Figure 17**.

Figure 18 shows sapphire fibers of 150–300 μm in diameter, which have been grown by multi-crystal process (up to 100 crystals per one process).

8. Conclusion

Comparable analysis of the forms and behaviors of the menisci profile curves for various signs of the boundary angle θ_d featuring the cases of ordinary (planar) and sloped working surfaces of the dies was made using the numerical analysis of the Young-Laplace capillary equation.

Calculation of the second derivatives of the menisci profile curves allowed us to analyze the curvature of the menisci profiles and observe its bulges.

Analysis of the inner and outer circular menisci has shown some different heights and behaviors of these menisci profiles because of the different signs of the azimuthal curvatures in the capillary equation.

Using numerical solution of the Young-Laplace capillary equation, the parameters for observing expression for automated growth control for tubular crystals as well as for capillaries and fibers were found. The sapphire tubes, capillaries, and fibers of good quality have been grown using automated system of control.

Acknowledgements

Numerical analysis of melt columns for fibers and growth of sapphire fibers was supported by the Russian Science Foundation (RSF), Project # 19-12-00402. Numerical analysis of menisci for capillaries was supported by the Russian Science Foundation (RSF), Project # 18-08-01230.

Author details

Sergei N. Rossolenko[1*], Gleb M. Katyba[1,2], Irina N. Dolganova[1,2], Irina A. Shikunova[1], Dmitry O. Stryukov[1], Kirill I. Zaitsev[2,3] and Vladimir N. Kurlov[1]

1 Institute of Solid State Physics RAS, Moscow, Russia

2 Bauman Moscow State Technical University, Moscow, Russia

3 Prokhorov General Physics Institute of the Russian Academy of Sciences, Moscow, Russia

*Address all correspondence to: ross@issp.ac.ru

References

[1] Sapphire KVN. Properties, growth, and applications. In: Hashmi S, editor. Reference Module in Materials Science and Materials Engineering. Oxford: Elsevier; 2016. pp. 1-11. DOI: 10.1016/B978-0-12-803581-8.03681-X

[2] Chalmers B, LaBelle HE, Mlavsky AI. Edge-defined, film-fed crystal growth. Journal of Crystal Growth. 1972;**13/14**:84-87. DOI: 10.1016/0022-0248(72)90067-X

[3] Stepanov AV. Future of Metal Working. Leningrad: Lenizdat; 1963. p. 130

[4] LaBelle HE. EFG, the invention and application to sapphire growth. Journal of Crystal Growth. 1980;**50**:8-17. DOI: 10.1016/0022-0248(80)90226-2

[5] Kurlov VN, Rossolenko SN, Abrosimov NV, Lebbou KH. Shaped crystal growth. In: Th D, editor. Crystal Growth Processes Based on Capillarity Czochralski, Floating Zone, Shaping and Crucible Techniques. Chichester, West Sussex, United Kingdom: John Wiley & Sons Ltd; 2010. pp. 277-354. DOI: 10.1002/9781444320237.ch5

[6] Antonov PI, Kurlov VN. A review of developments in shaped crystal growth of sapphire by the Stepanov and related techniques. Progress in Crystal Growth and Characterization of Materials. 2002;**44**:63-122. DOI: 10.1016/S0960-8974(02)00005-0

[7] Shikunova IA, Stryukov DO, Rossolenko SN, Kiselev AM, Kurlov VN. Neurosurgery contact handheld probe based on sapphire shaped crystal. Journal of Crystal Growth. 2017;**457**:265-269. DOI: 10.1016/j.jcrysgro.2016.08.062

[8] Zaytsev KI, Katyba GM, Kurlov VN, Shikunova IA, Karasik VE, Yurchenko SO. Terahertz photonic crystal waveguides based on sapphire shaped crystals. IEEE Transactions on Terahertz Science and Technology. 2016;**6**(4):576-582. DOI: 10.1109/TTHZ.2016.2555981

[9] Katyba GM, Zaytsev KI, Chernomyrdin NV, Shikunova IA, Komandin GA, Anzin VB, et al. Sapphire photonic crystal waveguides for terahertz sensing in aggressive environments. Advanced Optical Materials. 2018;**6**:1800573. DOI: 10.1002/adom.201800573

[10] Katyba GM, Zaytsev KI, Dolganova IN, Shikunova IA, Chernomyrdin NV, Yurchenko SO, et al. Sapphire shaped crystals for waveguiding, sensing and exposure applications. Progress in Crystal Growth and Characterization of Materials. 2018;**64**:133-151. DOI: 10.1016/j.perysgrow.2018.10.002

[11] Kurlov VN, Rossolenko SN. Growth of shaped sapphire crystals using automated weight control. Journal of Crystal Growth. 1997;**173**:417-426. DOI: 10.1016/S0022-0248(96)00836-6

[12] Abrosimov NV, Kurlov VN, Rossolenko SN. Automated control of Czochralski and shaped crystal growth processes using weighing techniques. Progress in Crystal Growth and Characterization of Materials. 2003;**46**:1-57. DOI: 10.1016/S0960-8974(03)90001-5

[13] Tsivinnski SV. Application of the theory of capillary phenomena to obtain products of a given shape directly from the melt by A.V. Stepanov method. Inzhenerno-fizichesky Zhurnal. 1962;5:59-65. (in Russian)

[14] Tatarchenko VA. Shaped Crystal Growth. Dordrecht: Kluwer Academic Publishers; 1993. p. 287

[15] Schelkin YF. Determining the shape of the column during the growth

of single crystals by the Czochralski method from a melt with a free surface. Fizika i Khimija Obrabotki Materialov. 1971;**3**:29-33. (in Russian)

[16] Surek T, Chalmers B, Mlavsky AI. The edge-defined film-fed growth of controlled shape crystals. Journal of Crystal Growth. 1977;**42**:453-465. DOI: 10.1016/0022-0248(77)90231-7

[17] Kuandykov LL, Antonov PI. Shaped melt column optimal choice on the basis of an equilibrium growth angle value. Journal of Crystal Growth. 2001;**222**:852-861. DOI: 10.1016/S0022-0248(00)01010-1

[18] Rossolenko SN. Menisci masses and weights in Stepanov (EFG) technique: Ribbon, rod, tube. Journal of Crystal Growth. 2001;**231**:306-315. DOI: 10.1016/S0022-0248(01)01448-8

[19] Satunkin GA, Rossolenko SN. Analysis of the dynamics of the controlled crystallization process using the Czochralski method. Crystal Research and Technology. 1986;**21**:1125-1138. DOI: 10.1002/crat.2170210902

[20] Red'kin BS, Kurlov VN, Pet'kov IS, Rossolenko SN. Investigation of the crystal growth conditions of gadolinium molybdate crystals. Journal of Crystal Growth. 1990;**104**:77-79. DOI: 10.1016/0022-0248(90)90310-H

[21] Rossolenko SN, Zhdanov AV. Equilibrium shapes of liquid menisci subjected to gravity force and surface tension. Journal of Crystal Growth. 1990;**104**:8-13. DOI: 10.1016/0022-0248(90)90300-A

[22] Landau LD, Lifshitz EM. Mehanika Splosnih Sred. Moscow: Gostehteorizdat; 1953. p. 788

[23] Satunkin GA. Determination of growth angles, wetting angles, interfacial tensions and capillary constant values of melts. Journal of Crystal Growth. 2003;**255**:170-189. DOI: 10.1016/S0022-0248(03)01187-4

[24] Ettouney HM, Brown RA, Kalejs JP. Analysis of operating limits in edge-defined film-fed crystal growth. Journal of Crystal Growth. 1983;**62**:230-246. DOI: 10.1016/0022-0248(83)90301-9

[25] Thomas PD, Ettouney HM, Brown RA. A thermal-capillary mechanism for a growth rate limit in edge-defined film-fed growth of silicon sheets. Journal of Crystal Growth. 1986;**76**:339-351. DOI: 10.1016/0022-0248(86)90380-5

[26] Thomas PD, Brown RA. Rate limits in silicon sheet growth: The connections between vertical and horizontal methods. Journal of Crystal Growth. 1987;**82**:1-9. DOI: 10.1016/0022-0248(87)90157-6

[27] Derby JJ, Brown RA. On the dynamics of Czochralski crystal growth. Journal of Crystal Growth. 1987;**83**:137-151. DOI: 10.1016/0022-0248(87)90514-8

[28] Lan CW. Thermal-capillary analysis of floating-zone growth of tube crystals: Steady-state and condition dominated calculations. Journal of Crystal Growth. 1994;**135**:606-618. DOI: 10.1016/0022-0248(94)90154-6

[29] Yeckel A, Salinger AG, Derby JJ. Theoretical analysis and design considerations for float-zone refinement of electronic grade silicon sheets. Journal of Crystal Growth. 1995;**152**:51-64. DOI: 10.1016/0022-0248(95)00088-7

[30] Roy A, Zhang H, Prasad V, Mackintosh B, Quellette M, Kalejs JP. Growth of large diameter silicon tube by EFG technique: Modeling and experiment. Journal of Crystal Growth. 2001;**230**:224-231. DOI: 10.1016/S0022-0248(01)01327-6

[31] Satunkin GA, Tatarchenko VA. Shape analysis and meniscus height calculations for various types of capillary shaping. Journal of Colloid and Interface Science. 1985;**104**:318-333. DOI: 10.1016/0021-9797(85)90042-6

[32] Zhdanov AV, Satunkin GA, Ponomareva RP. Stability of liquid menisci. Journal of Colloid and Interface Science. 1985;**104**:334-343. DOI: 10.1016/0021-9797(85)90043-8

[33] Kurlov VN, Epelbaum BM. Fabrication of near-net-shaped sapphire domes by noncapillary shaping method. Journal of Crystal Growth. 1997;**179**:175-180. DOI: 10.1016/S0022-0248(97)00111-5

Chapter 4

Growth of Single-Crystal $LiNbO_3$ Particles by Aerosol-Assisted Chemical Vapor Deposition Method

José G. Murillo, José A. Ocón, Guillermo M. Herrera, José R. Murillo-Ochoa and Gabriela Ocón

Abstract

Adjusting nucleation conditions, an effective shape and size control in the preparation of single-crystal lithium niobate nanoparticles by aerosol-assisted chemical vapor deposition method was demonstrated. The effect of the most relevant parameters leading to nanocrystals taking a specific shape or size once they are synthesized was analyzed. This has allowed us to demonstrate that it is possible to control the size and morphology of particles prepared adjusting the nucleation conditions. The synthesized nanocrystals showed different morphologies including quasi-cubic, tetrahedral, polyhedral, and hexagonal shapes, with characteristic sizes ranging from a few tens to a few hundred nanometers. However, rod-like structures with characteristic lengths ranging from 3 to 5 μm were also obtained. The structural and morphological characterization by X-ray diffraction and high-resolution electron microscopy techniques revealed the single-crystal nature of the synthesized particles.

Keywords: $LiNbO_3$ nanocrystals, aerosol-assisted chemical vapor deposition, nanomaterials, niobates, ferroelectric materials

1. Introduction

Lithium niobate ($LiNbO_3$) has been one of the most prominent and widely studied ferroelectric materials in the last decades. Their extraordinary pyroelectric, piezoelectric, and strong nonlinear optical properties have turned it into a key material in photonics and integrated optics [1–4]. The diverse, large-magnitude physical properties of $LiNbO_3$ (LN) have allowed it to be used in information storage and processing [5], in optical channel waveguides [6, 7], and in electro-optical modulation [8].

Moreover, in the last decade, a great interest has arisen for the synthesis of nanosized LN, including nanowires and nanoparticles [9–15]. This growing interest in the synthesis of nanoscale materials is originated by their new properties with potential technological application arising from the dimensional confinement [16, 17]. In view of that, in this research we describe in detail the synthesis and the morphological and structural characterization of LN nanoparticles using high-resolution electron microscopy techniques.

To obtain LN particles, several synthesis routes have been reported in literature, which have resulted in different shapes, sizes, and degree of crystallinity. Flake-shaped LN nanocrystals with 40–100 nm in diameter and LN anisotropic nanorods with lengths up to 100 and 7 nm diameters have already been obtained by hydrothermal route and by solution-phase synthesis [18–20]. By coprecipitation method LN particles have been synthesized with a loosely porous packed shape [21]. Furthermore, using sol-gel method, powders that after thermal treatment present the crystalline phase of LN with sizes lower than 100 nm and a roughly spherical geometry have been synthesized [22]. Under solvothermal conditions cubic and sphere-like LN nanoparticles have been prepared [23]. Nevertheless, there are still some important challenges to be solved such as growth direction, size and shape control, and degree of crystallinity.

The aim of this research is to describe in detail the synthesis process of LN nanocrystals, to present their morphology and structural characteristics, and also to demonstrate that it is possible to control the size and morphology of LN nanocrystals synthesized by the aerosol-assisted chemical vapor deposition method (AACVD). We describe a detailed characterization of the LN particles synthesized, including an analysis of the nucleation conditions that allow the control of their size and morphology. We have analyzed the effect of the most relevant parameters leading to the nanocrystals taking a specific shape or size when synthesized by the AACVD method. It was observed that depending on the exact conditions of the synthesis process, nanocrystals showed different morphologies including quasi-cubic, tetrahedral, polyhedral, and hexagonal shapes, with characteristic sizes ranging from a few tens to a few hundred nanometers or even rod-like particles with characteristic lengths ranging from 3 to 5 μm. The nanocrystals prepared were characterized by grazing incidence X-ray diffraction (GIXRD), scanning electron microscopy (SEM), and high-resolution transmission electron microscopy (HRTEM).

2. Experimental procedure

The LN nanocrystals were prepared following the same procedure used to obtain a thin film by the AACVD method as described in Refs. [24, 25]. That is, a silicon (0 0 1) or sapphire (1 1 0) substrate (1.5 × 1.5 cm^2) was used as a support medium for nanoparticle growth. In the LN nanoparticles' synthesis process, the determination of the deposition conditions was based on the route used in Ref. [26] to prepare a LN thin film. Nevertheless, some of the nucleation conditions useful for growing LN thin films were modified or optimized to obtain LN nanocrystals. Silicon was selected as substrate in order to combine its electronic advantages with the LN piezoelectric properties and optical processing capabilities. Concerning the sapphire substrate, its lattice parameters similar to LN enable the hetero-epitaxial growth of LN piezoelectric particles, i.e., with their atomic planes built following the same crystallographic orientation of the atomic planes on the substrate surface [26].

In order to evaluate possible changes in the crystalline structure, size, and morphological characteristics of the LN particles synthesized, two deposition systems were used, one with a fixed and the other with a mobile nozzle. In both cases the nozzle discharge surface was 1 mm × 150 mm.

The fixed nozzle deposition system uses an ultrasonic nebulizer working at 2.4 MHz to generate the aerosol precursor solution which is conveyed by a carrier gas (dry and clean air) and directed toward the substrate by the nozzle. The substrate is directly in contact with a metallic plate heated at the selected temperature. The general properties and details of the deposition systems are published elsewhere [25]. The nanocrystals were prepared at different temperatures covering the range from 653 to 783 K. In order to determine the optimal conditions to obtain LN nanocrystals keeping control of the size and shape, we also varied the carrier gas flow, the nozzle-to-substrate distance, and deposition time. The other deposition system is very similar to the system described above, except for the use of a mobile nozzle which describes a back-and-forth movement at constant velocity (~ 1 cm min^{-1}) to scan the whole surface of the substrate. By varying the total number of scans, we obtained particles of different sizes. The starting solutions for feeding the ultrasonic nebulizer which generates the aerosol precursor were dilutions of niobium ethoxide (99.95% Sigma-Aldrich) and lithium acetylacetonate (97% Sigma-Aldrich) in methanol or ethanol, using concentrations from 0.01 to 0.1 mol l^{-1}.

The morphological and structural characterization of the LN nanocrystals obtained was completed by GIXRD, SEM, and HRTEM. The crystalline phases present in nanoparticles were determined by GIXRD patterns in a Panalytical X-Pert system using Cu Kα radiation (λ = 0.15418 nm) at 40 keV and 35 mA. The grazing incidence angle was fixed at 0.5°, and the scanning angle 2θ was varied from 20 to 80° at 0.1°min^{-1} intervals. The surface morphology was studied by field emission SEM using a JEOL JSM-7401F operated at 2 kV.

3. Results and discussion

3.1 Synthesis

We accomplished a systematic variation of the nucleation conditions used in nanocrystals growth process, including substrate type and temperature, carrier gas flux of the precursor aerosol, nozzle-to-substrate distance, and solvent type (methanol or ethanol) used to prepare the precursor solution.

The LN nanocrystals obtained showed different sizes and morphologies including multifaceted shapes, quasi-cubic shapes, tetrahedral shapes, diamond-like polyhedrons, hexagonal prism shapes, or hexagonal rod-like shapes depending on the exact nucleation conditions used in the synthesis process. In fact, a large number of deposition cycles were carried out by AACVD method on silicon (0 0 1) or sapphire (1 1 0) substrates, to investigate the possibility of obtaining LN particles with different shapes and sizes. Some representative LN structures synthesized on silicon or sapphire substrates with both the fixed nozzle system and the mobile nozzle system are shown below. **Figure 1(a)–(c)** shows SEM micrographs describing the morphology of some typical LN nanocrystals obtained using ethanol as solvent in the precursor solution. **Figure 1(d)–(h)** shows SEM micrographs of the morphology of some LN nanocrystals obtained dissolving precursor solution in methanol.

3.2 Morphological and structural characterization

It was found from a detailed analysis of the controlled nucleation conditions used in the crystalline structure growth process that substrate temperature is a particularly crucial parameter in the synthesis process. The optimal temperature range required to obtain LN crystals with faceted shapes was 653–783 K. Temperatures out of this range caused the formation of non-faceted particles with rock-like or island-like shapes similar to a piece of thin film. The carrier gas flow magnitude was also decisive in the shape of LN crystals. Indeed, fluxes between 5 and 6 l/min led to the

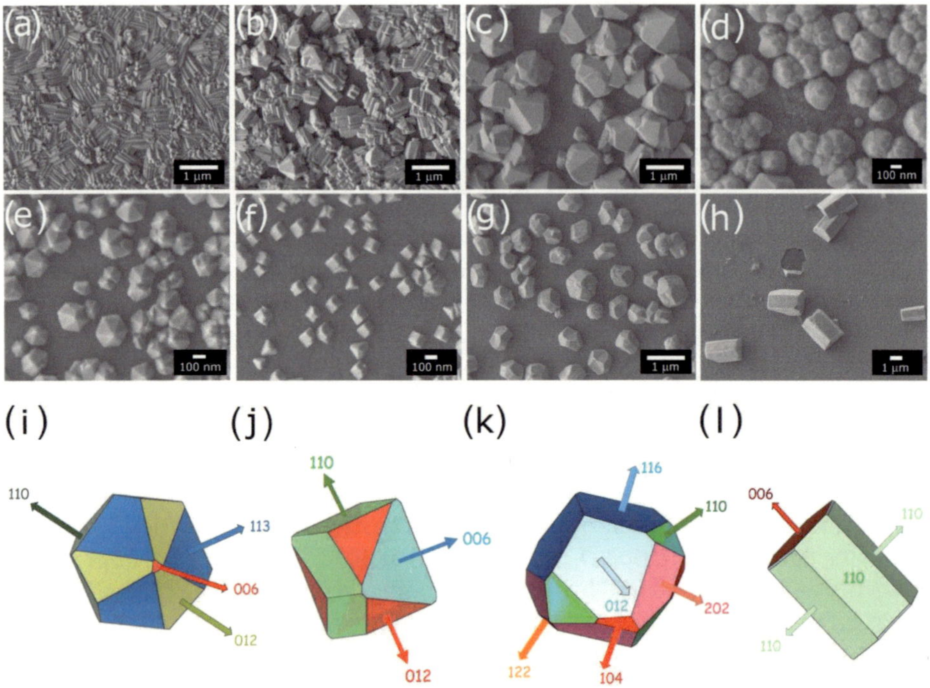

Figure 1.
(a)–(c) Typical LN nanocrystals obtained using ethanol as solvent in the precursor solution, and prepared on a Si substrate, (a) and (b) rod-like type I particles, (c) polyhedral type II particles. (d)-(h) Representative LN nanocrystals obtained dissolving the precursor solution in methanol. (d), (e), (g) and (h) prepared on a (001) Si substrate, and (f) on a (1 1 0) sapphire substrate. (d) Agglomerated polyhedral type III particles, (e) diamond-like polyhedral type IV particles, (f) a mix of quasi-cubic type V and tetrahedral type VI nanoparticles, (g) polyhedral type VII particles, and (h) hexagonal prism type VIII particles. (i), (j), (k) and (l) particles morphology showing the crystallographic orientation of the facets in selected particles of representative samples LN73S, LN37SP, LN53S and LN128S shown in Figure (e), (f), (g) and (h), respectively, reproduced using the software VESTA[29] simulated according to the texture observed in the corresponding GIXRD results.

formation of rod-like particles (**Figure 1(a)** and **(b)**), while fluxes between 2 and 3 l/min led to faceted particles with polyhedral morphology or diamond-like shape (**Figure 1(e)–(g)**). The molar concentration of the precursors diluted in methanol or ethanol also had an important effect on the LN crystal shape and size. The use of low molar precursor concentrations in the solution to be sprayed of the order of 0.01–0.02 mol l^{-1} led to the formation of polyhedral diamond-like particles at 773 K (**Figure 1(e)**), hexagonal prism shapes at 653 K (**Figure 1(h)**), or quasi-cubic shapes if the substrate is sapphire instead of silicon at 723 K (**Figure 1(f)**).

On the other hand, the use of methanol or ethanol as solvent in the precursor solution used in the synthesis process had a strong effect on the determination of LN particle morphology. This finding is in agreement with the results reported in Ref. [27] where the great influence of alcohols used as solvents on the morphology and size of iron oxide nanoparticles synthesized by sol–gel method is evinced. As shown in **Figure 1**, we obtained rod-like type I particles (**Figure 1(a)–(b)** samples LN11S and LN12S), polyhedral type II particles (**Figure 1(c)** sample LN30S), agglomerated polyhedral shape type III particles (**Figure 1(d)** sample LN77S), polyhedral diamond-like type IV particles (**Figure 1(e)** sample LN73S), quasi-cubic type V and tetrahedral type VI particles (**Figure 1(f)** sample LN37SP), polyhedral type VII particles (**Figure 1(g)** sample LN53S), or even hexagonal prism type VIII particles (**Figure 1(h)** sample LN128S). Specifically, it was found that the use of methanol to dilute the precursor solution enables the formation of polyhedral type III, diamond-like polyhedral (type IV), quasi-cubic type V, tetrahedral type VI, and polyhedral type VII particles at the adequate temperature (653–773 K) (**Figure 1(d)–(g)**). However, the use of ethanol as solvent in the precursor solution enables the formation of extended rod-like particles and polyhedral type II particles, using the fixed nozzle deposition system (**Figure 1(a)–(c)**). Nevertheless, using methanol in the dilution of precursor and the mobile nozzle system, it is also possible to obtain hexagonal prism type VIII particles (**Figure 1(h)**). In addition to this, the specific characteristics of the synthesized LN particles also depended on the specific deposition system used. The fixed nozzle deposition system enables the production of rod-like particles (**Figure 1(a)–(b)**) and polyhedral type II particles (**Figure 1(c)**), due to the high amount of material sprayed into the substrate. In contrast, the mobile nozzle system enables the production of polyhedral and diamond-like particles (**Figure 1(c)–(e)**). Moreover, in **Figure 1(f)** it can be observed that it is possible to produce particles with quasi-cubic and tetrahedral morphologies, keeping the nucleation parameters close to the required values and changing only the type of substrate. It is clear then that the substrate crystallographic orientation plays an important role in the resultant morphology of synthesized LN particles, since the crystallographic orientation of the sapphire substrate (1 1 0) was different to that of the silicon substrate (0 0 1) used. LN particles in sample LN37SP (**Figure 1(f)**) were synthesized on a sapphire substrate using a set of nucleation parameters very similar to those used to produce the particles in sample LN53S (**Figure 1(g)**). Furthermore, it was observed that under certain nucleation conditions, the synthesis of nanocrystals could finish in a continuous layer or thin film. **Figure 1(h)** shows six very large LN hexagonal structures over a continuous LN thin film obtained at a temperature of 653 K with the mobile nozzle system, using a carrier gas flow of 1 l/min, and a deposition time of 130 min. **Figure 1(i)–(l)** shows the particle morphology of samples LN73S, LN37SP, LN53S, and LN128S shown, respectively, in **Figure 1(e)–(h)** reproduced using the software VESTA [28] that additionally allowed us to determine the crystallographic orientation of these particle facets. The crystallographic facets of the particles in these figures were simulated according to the texture observed in their corresponding GIXRD results. **Table 1** summarizes the nucleation conditions used to obtain some of the representative LN particles studied in this work.

The crystalline nature of some of the representative particles was verified from its GIXRD pattern. **Figure 2(a)–(d)** shows the comparison between GIXRD patterns for samples prepared with methanol labeled as LN53S, LN37SP, and LN73S with respect to the sample prepared with ethanol labeled as LN11S. The crystalline structure for these GIXRD patterns was related to lithium niobate, and the diffraction peaks were indexed through the JCPD 01-070-8451 file [29]. These results correspond to a rhombohedral phase with a R3c (161) space group. Other diffraction peaks labeled with asterisk correspond to the Si substrate, JCPD 00-027-1402, and those with a cross to the sapphire substrate, JCPD 00-001-1305 [29]. The GIXRD pattern for the samples prepared with methanol showed two diffraction peaks denoted by (012) and (104). This result suggests the presence of a preferred growth orientation (texture) with respect to the sample which was prepared with ethanol. The presence of texture in sample LN73S prepared with methanol could be attributed to a combination of synthesis parameters during the sample preparation such as temperature, molar concentration, growth time, type of solvent, fixed or mobile nozzle system, air flux, etc. Furthermore, it can be seen in **Figure 2(c)** that the diffraction peaks in the GIXRD pattern in sample LN73S show a sharper and narrower profile than the diffraction peaks in the GIXRD pattern of sample LN11S shown in **Figure 2(d)**.

Sample	Substrate	Solvent	Precursor concentration (mol/l)	Substrate temperature (K)	Carrier gas flow (l/min)	Nozzle system	Scanning rate system (cm/min)	Growth time (min)
LN11S	Silicon	Ethanol	0.1	783	5	Fixed	—	30
LN12S	Silicon	Ethanol	0.1	773	5	Fixed	—	30
LN30S	Silicon	Ethanol	0.05	773	5	Mobile	0.2	120
LN37SP	Sapphire	Methanol	0.02	723	2	Fixed	—	60
LN53S	Silicon	Methanol	0.02	723	2	Fixed	—	120
LN73S	Silicon	Methanol	0.02	723	3	Mobile	1	60
LN77S	Silicon	Methanol	0.01	773	3	Mobile	1	40
LN128S	Silicon	Methanol	0.01	653	1	Mobile	1	130

Table 1.
Nucleation conditions used to synthesize representative LN particles.

DOI: http://dx.doi.org/10.5772/intechopen.88323

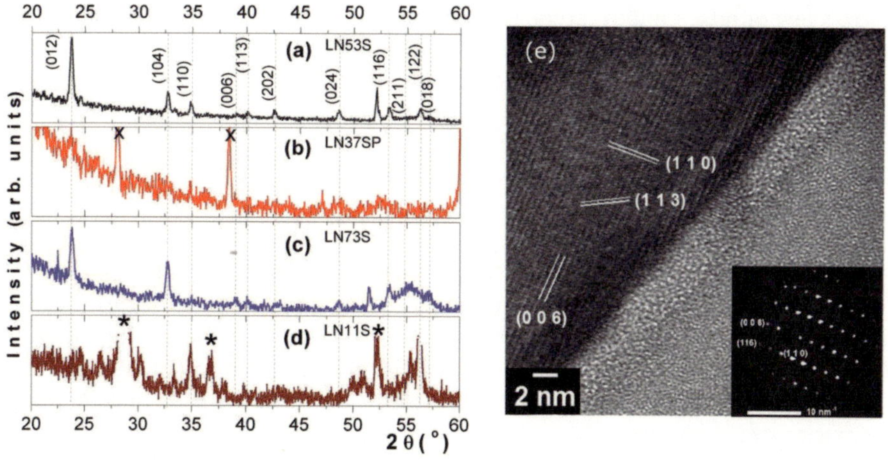

Figure 2.
*Comparison between GIXRD patterns for samples prepared with methanol labeled as (a) LN53C, (b) LN37SP, and (c) LN73S, respect to the sample prepared with ethanol labeled as (d) LN11S, other diffraction peaks labeled with asterisk correspond to the Si substrate and those with a cross to the sapphire substrate. (e) Bright field HRTM image of the interface between a LN nanocrystal such as shown in **Figure 1(e)** and a Si substrate showing the interplanar distance of the (0 0 6), (1 1 0) and (1 1 3) family planes of the lithium niobate hexagonal phase. In the lower half of the micrograph is shown a sharp interface between a $LiNbO_3$ nanocrystal and an amorphous native thin silicon oxide interlayer. The inset in the lower right corner shows a selected area electron diffraction pattern demonstrating the existence of the single crystal hexagonal phase.*

In order to determine the particle mean crystal size in sample LN73S, we used the Scherrer equation:

$$C = 0.9\,\lambda/\beta\cos\theta \qquad (1)$$

where C is the mean crystal size, λ is the wavelength of CuKα (1.54 Å), β is the full width half maximum (FWHM) intensity corrected using LaB6 as the standard, and θ is Bragg's angle of the selected peaks (=23.79 and 32.81°). The mean crystal size for particle in LN73S sample was 33.9 nm which matches the dimensions shown in the morphology micrographs evaluated by SEM.

GIXRD results were confirmed by HRTEM techniques preparing first a sample with a cross-section of some LN nanocrystals as those shown in **Figure 1(e)**. **Figure 2(e)** shows a bright-field high-resolution TEM image of the interface between the Si substrate and a LN nanocrystal. The lattice fringe measurement (**Figure 2(e)**) was consistent with the interplanar distance of the (0 0 6), (1 1 0), and (1 1 3) family planes of the lithium niobate hexagonal phase [29]. The inset in **Figure 2(e)** shows a selected area electron diffraction pattern which confirms the existence of the hexagonal phase and the single-crystal nature of the LN particle shown in **Figure 1(e)**.

4. Conclusions

We described in detail the synthesis process of LN nanocrystals prepared by AACVD method, and we analyzed the effect of the most relevant parameters leading to nanocrystals taking a specific shape or size once they are synthesized. This has allowed us to demonstrate that it is possible to control the size and morphology of particles prepared adjusting the nucleation conditions. The synthesized nanocrystals showed different morphologies including quasi-cubic, tetrahedral, polyhedral, and hexagonal shapes, with characteristic sizes ranging from a few tens to a few hundred nanometers. In addition to this, we also obtained rod-like structures with characteristic lengths ranging from 3 to 5 μm. The characterization of the prepared LN particles by GIXRD, SEM, and HRTEM techniques revealed the single-crystal nature of the synthesized particles.

Acknowledgements

The authors want to thank W. Antúnez-Flores, O. Solís, C. Ornelas-Gutiérrez, E. Guerrero, and J.T. Holguin, for experimental assistance, the Laboratorio Nacional de Nanotecnología and Centro de Investigación en Materiales Avanzados S.C. (CIMAV) for the facilities provided, and Ivan Rodríguez (Facultad de Filosofía y Letras, UACh, Universidad Autónoma de Chihuahua) for the assistance in the preparation of the chapter.

Conflict of interest

The authors declare no conflict of interest.

Abbreviations

LN	lithium niobate
AACVD	aerosol-assisted chemical vapor deposition
GIXRD	grazing incidence X-ray diffraction
SEM	scanning electron microscopy
HRTEM	high-resolution transmission electron microscopy

Author details

José G. Murillo[1]*, José A. Ocón[2], Guillermo M. Herrera[3], José R. Murillo-Ochoa[4] and Gabriela Ocón[5]

1 Departmento de Física de Materiales, Centro de Investigación en Materiales Avanzados S. C., Chihuahua, Mexico

2 Universidad Tecnológica de Chihuahua, Chihuahua, Mexico

3 Cátedra CONACYT asignada al Departmento de Física de Materiales, Centro de Investigación en Materiales Avanzados S. C., Chihuahua Mexico

4 Universidad del Valle de Mexico, Chihuahua, Mexico

5 Universiadad Autónoma de Chihuahua, Chihuahua, Mexico

*Address all correspondence to: jose.murillo@cimav.edu.mx

References

[1] Arizmendi L. Photonic applications of lithium niobate crystals. Physica Status Solidi A: Applied Research. 2004;**201**:253-283

[2] Yim YS, Shin SY. Lithium niobate integrated-optic voltage sensor. Optics Communication. 1998;**152**:225-228

[3] Guarino A, Poberaj G, Rezzonico D, Degl'Innocenti R, Günter P. Electro-optically tunable microring resonators in lithium niobate. Nature Photonics. 2007;**1**:407-410

[4] Murillo JG, Herrera G, Vega-Rios A, Flores-Gallardo S, Duarte-Moller A, Castillo-Torres J. Effect of Zn doping on the photoluminescence properties of $LiNbO_3$ single crystals. Optical Materials. 2016;**62**:639-645

[5] Buse K, Adibi A, Psaltis D. Non-volatile holographic storage in doubly doped lithium niobate crystals. Nature. 1998;**393**:665-668

[6] Holm A, Stürzer Q, Xu Y, Weigel R. Investigation of surface acoustic waves on $LiNbO_3$, quartz, and $LiTaO_3$ by laser probing. Microelectronic Engineering. 1996;**31**:123-127

[7] Chang SJ, Tsai CL, Lin YB, Liu JF, Wang WS. Improved electrooptic modulator with ridge structure in x-cut $LiNbO_3$. Journal of Lightwave Technology. 1999;**17**:843-847

[8] Wooten EL, Kissa KM, Yi-Yan A, Murphy EJ, Lafaw DA, Hallemeier PF, et al. A review of lithium niobate modulators for fiber-optic communications systems. IEEE Journal of Selected Topics in Quantum Electronics. 2000;**6**:69-82

[9] Zhao L, Steinhart M, Yosef M, Lee SK, Schlecht S. Large-scale template-assisted growth of $LiNbO_3$ one-dimensional nanostructures for nano-sensors. Sensors and Actuators B: Chemical. 2005;**109**:86-90

[10] Guttmann E, Benke A, Gerth K, Böttcher H, Mehner E, Klein C, et al. Pyroelectrocatalytic disinfection using the pyroelectric effect of nano- and microcrystalline $LiNbO_3$ and $LiTaO_3$ particles. Journal of Physical Chemistry C. 2012;**116**:5383-5393

[11] Aufray M, Menuel S, Fort Y, Eschbach J, Rouxel D, Vincent B. New synthesis of nanosized niobium oxides and lithium niobate particles and their characterization by XPS analysis. Journal of Nanoscience and Nanotechnology. 2009;**9**:4780-4785

[12] Santulli AC, Zhou H, Berweger S, Raschke MB, Sutter E, Wong SS. Synthesis of single-crystalline one-dimensional $LiNbO_3$ nanowires. CrystEngComm. 2010;**12**:2675-2678

[13] Grigas AS, Kaskel S. Synthesis of $LiNbO_3$ nanoparticles in a mesoporous matrix. Beilstein Journal of Nanotechnology. 2011;**2**:28-33

[14] Cheng LQ, Li JF. A review on one dimensional perovskite nanocrystals for piezoelectric applications. Journal of Materiomics. 2016;**2**:25-36

[15] Boltersdorf J, King N, Maggard PA. Flux-mediated crystal growth of metal oxides: Synthetic tunability of particle morphologies, sizes, and surface features for photocatalysis research. CrystEngComm. 2015;**17**:2225-2241

[16] Wu N, Wang J, Tafen D, Wang H, Zheng J, Lewis JP, et al. Shape-enhanced photocatalytic activity of single-crystalline anatase TiO_2 (101) nanobelts. Journal of the American Chemical Society. 2010;**132**:6679-6685

[17] Zhu S, Zhang J, Qiao C, Tang S, Li Y, Yuan W, et al. Strongly

green-photoluminescent graphene quantum dots for bioimaging applications. Chemical Communications. 2011;**47**:6858-6860

[18] An C, Tang K, Wang C, Shen G, Jin Y, Qian Y. Characterization of $LiNbO_3$ nanocrystals prepared via a convenient hydrothermal route. Materials Research Bulletin. 2002;**37**:1791-1796

[19] Grange R, Choi JW, Hsieh CL, Pu Y, Magrez A, Smajda R, et al. Lithium niobate nanowires synthesis, optical properties, and manipulation. Applied Physics Letters. 2009;**95**:143105

[20] Wood BD, Mocanu V, Gates BD. Solution-phase synthesis of crystalline lithium niobate nanostructures. Advanced Materials. 2008;**20**:4552-4556

[21] Prakash BJ, Buddhudu S. Synthesis and characterization of Mn^{2+}: $LiNbO_3$ nano-particles by co-precipitation method. Indian Journal of Pure and Applied Physics. 2013;**51**:245-250

[22] Graça MPF, Prezas PR, Costa MM, Valente MA. Structural and dielectric characterization of $LiNbO_3$ nano-size powders obtained by Pechini method. Journal of Sol-Gel Science and Technology. 2012;**64**:78-85

[23] Mohanty D, Chaubey GS, Ourdkhani A, Adireddy S, Caruntu G, Wiley JB. Synthesis and piezoelectric response of cubic and spherical $LiNbO_3$ nanocrystals. RSC Advances. 2012;**2**:1913-1916

[24] Ocón JA, Murillo JG, Miki-Yoshida M, Cardoza MN, Contreras-López OE. Synthesis and characterization of $LiNbO_3$ nanocrystals prepared by the aerosol assisted chemical vapor deposition method. Journal of Crystal Growth. 2014;**408**:64-70

[25] Amezaga-Madrid P, Antunez-Flores W, Ledezma-Sillas JE, Murillo-Ramírez JG, Solís-Canto O, Vega-Becerra OE, et al. Synthesis, microstructural characterization and optical properties of undoped, V and Sc doped ZnO thin films. Journal of Alloys and Compounds. 2011;**509S**:S490

[26] Bornand V, Huet I, Bardeau JF, Chateigner D, Papet P. An alternative route for the synthesis of oriented $LiNbO_3$ thin films. Integrated Ferroelectrics. 2002;**43**:51-64

[27] Joshi DP, Pant G, Arora N, Nainwal S. Effect of solvents on morphology, magnetic and dielectric properties of (α-Fe_2O_3@SiO_2) core-shell nanoparticles. Heliyon. 2017;**3**:e00253

[28] Momma K, Izumi F. VESTA 3 for three-dimensional visualization of crystal, volumetric and morphology data. Journal of Applied Crystallography. 2011;**44**:1272-1276

[29] Joint Committee on Powder Diffraction Standards. Powder Diffraction File. card 01-070-8451. Swarthmore, PA: International Center for Diffraction Data; 2006

Section 3

Usual and Unusual Growth

Chapter 5

Epitaxial Growth of Thin Films

Daniel Rasic and Jagdish Narayan

Abstract

Epitaxial thin film heterostructures are critical for integrating multi-functionality on a chip and creating smart structures for next-generation solid-state devices. Here, we discuss the traditional lattice matching epitaxy (LME) for small lattice misfit and domain matching epitaxy (DME), which handles epitaxial growth across the misfit scale, where lattice misfit strain is predominant and can be relaxed completely, meaning that only the thermal and defect strains remain upon cooling. In low misfit systems, all three sources contribute to the residual strain upon cooling, as result of incomplete lattice relaxation. In the second part of the chapter, we will discuss the two critical contributors to the stress of the epitaxial film: the thermal coefficient of expansion mismatch and the lattice plane misfit. In the last part of the chapter, we will focus on unique cases where room temperature epitaxial growth is possible in nitride and oxide thin films.

Keywords: thin films, annealing, defects, epitaxy, lattice misfit

1. Introduction

The modern civilization operates in bits (zeroes and ones), and the start of the "binary" (digital) era was made possible by the invention of transistor in 1947. Since then, the improvements were growing at exponential rate by halving the size of each transistor and doubling the processing speed each year (a.k.a. Moore's law) [1]. The fundamental template on which transistors are made is silicon. Experimentally, the modern transistors are made using thin-film growth technologies. With each size decrease, new scaling issues occurred primarily because of deposition technology limitations of that era. However, as we move closer to single nanometer nodes, fundamental limitations that originate from material properties start to take over as the main challenges that were previously reserved for instrumentation [2, 3].

In this chapter, the overview of thin-film growth is provided, followed by the discussion on epitaxy and lattice misfit considerations. Finally, the role of temperature in film growth is discussed with some examples.

1.1 Pulsed laser deposition

One of the most versatile and powerful growth methods used in thin-film growth used today is pulsed laser deposition (PLD). Main advantages over sputtering, physical vapor deposition (PVD), chemical vapor deposition (CVD) and other techniques is its forward-directed, non-equilibrium growth of multi-layered novel and exciting materials that preserve the target stoichiometry at much lower temperatures [4–7].

Perhaps the main advantage of the PLD process is the ability to deposit several multi-component materials in-situ with stoichiometry preserved [7]; an impossible feat with equilibrium methods. In **Figure 1,** a schematic of a PLD assembly is shown. The high-energy pulsed laser is used to ablate the target and form a plasma-like plume that deposits target material on the substrate of choice. The number of laser pulses during growth dictates the thickness of the film, and rotating the target assembly allows for multi-layer growth. In PLD, the energy of the ablated species is ~16 eV, compared to the energy of the ejected species in electron and thermal beam evaporation techniques (e.g. thermal energy, kT at 1200 K is ~0.1 eV). The high-energy of deposited species lowers the overall thermal budget requirement for epitaxial growth by providing enhanced mobility and growth on the substrates [8].

Epitaxy describes the oriented growth of a crystalline material on top of another. In a more specific term, it describes, predominately, the thin-film growth of a material on a substrate with a well-defined relationship. In the case of substrate/film growth, the film and the substrate have different chemistry, crystallographic properties and expansion coefficients. Therefore, most substrate/film combinations will not yield epitaxial films. The most obvious, first-order consideration to determine the probability of epitaxial growth in substrate/film growth is by calculating the lattice misfit at the interface by

$$\varepsilon_m(\%) = \left(1 - \frac{d_f}{d_s}\right) \times 100 \tag{1}$$

where d_f and d_s are lattice plane spacing in the in-plane directions between the film and the substrate, respectively.

Epitaxial thin film growth is almost always preferred over textured (single fixed axis alignment between substrate and film) or polycrystalline growth because the physical properties; e.g. conductivity can be enhanced and controlled [9]. Epitaxial thin films also need considerably smaller amounts of material to achieve performance comparable to randomly oriented (polycrystalline or nanocrystalline) films, which becomes an important consideration when using toxic or rare materials.

1.2 Lattice matching epitaxy

When the lattice misfit is small (7% > ε_m > 0%), the film growth occurs with one-to-one matching of lattice planes. Due to the difference in the lattice constants, lattice

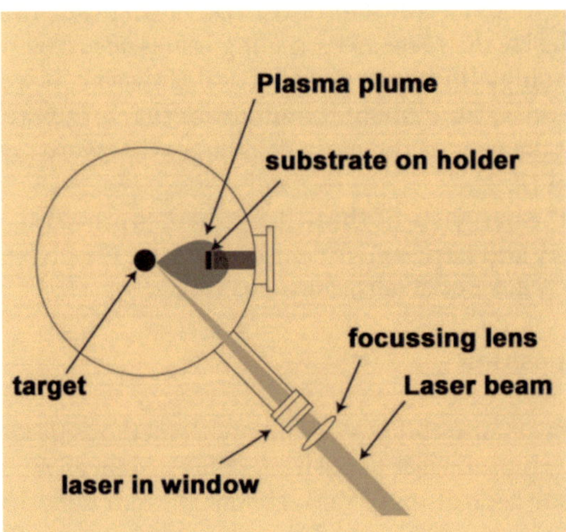

Figure 1.
Sketch showing the pulsed laser deposition assembly.

strain in the film is created that increases with film thickness. The accumulated strain is subsequently relaxed, at a certain thickness, by dislocation nucleation on the surface and propagation to the interface of the film and the substrate. The thickness at which this occurs is called the critical thickness of the film and is dictated by the lattice misfit, nucleation energy of dislocation and available slip systems as determined by the Schmidt factor. This means that the critical thickness is inversely correlated to the lattice misfit, i.e. the larger lattice misfit will cause relaxation via dislocations at a smaller film thickness.

1.3 Domain matching epitaxy

It was shown in recent times that the traditional lattice matching epitaxy (LME) is not able to explain the growth of several substrate/film heterostructures. For example, epitaxial titanium nitride thin films were grown on silicon by Narayan et al. as shown in **Figure 2** [10]. In the mentioned case, the lattice misfit is ~22.5%, suggesting that the lattice strain that is created on the film by the substrate would cause an immediate dislocation at the first step of film growth. To address this large misfit epitaxial thin film growth, the concept of domain-matching epitaxy (DME) was introduced that provides a universal paradigm for thin film epitaxy across the misfit scale [11]. The DME paradigm is centered around matching of integral multiples of lattice planes while the misfit in-between the integral multiples is relieved by the principle of domain variation, where alternation of domains occurs with a certain frequency to accommodate the misfit.

Theoretically, if the misfit yields the perfect matching ratios of planes ($md_f = nd_s$), the residual strain will be zero. Contrastingly, if the misfit yields non-zero residual strain, it is alleviated by two domains, alternating with a certain frequency (α) to minimize the residual strain, given as

$$(m + \alpha)\, d_f = (n + \alpha)\, d_s \tag{2}$$

where α is the domain variation (frequency) parameter. For example, if $\alpha = 0.5$, then m/n and (m + 1)/(n + 1) domains alternate with the same frequency [11] With

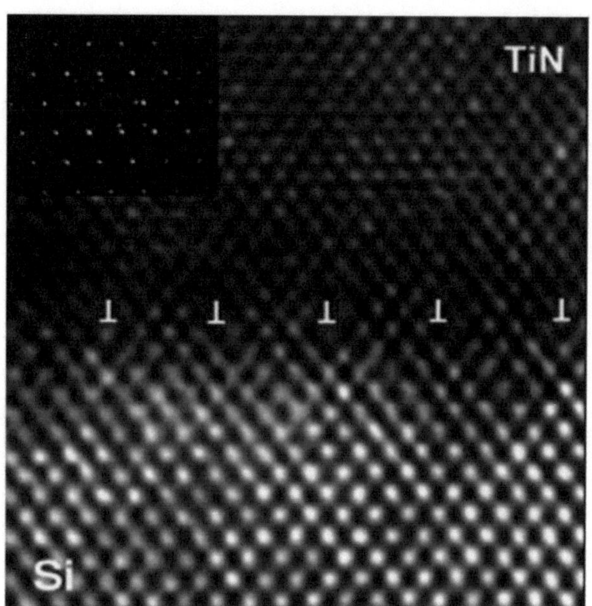

Figure 2.
HRTEM image of TiN/Si(1 0 0) interface with alternating 4/3 and 5/4 domains [10].

additional research, it was shown that epitaxial growth is possible for many film/substrate systems such as AlN/Al_2O_3, TiN/Al_2O_3, ZnO/Al_2O_3, MgO/Al_2O_3, MgO/STO, LSMO/MgO, MgO/Si, STO/Si, etc. [11, 12]. The two critical steps in the kinetics of thin film relaxation and growth are dislocation nucleation and subsequent propagation. Firstly, nucleation of dislocations occurs at the free surface, and the nucleation barrier is determined by the presence of surface steps. Experimentally, it was demonstrated that the nucleation of dislocations is lower when the film is under compressive stress as opposed to tensile stress [13]. This concept can be applied in the case of Si/Ge system where the critical thickness is found to be larger for films under tensile stress than in the Ge/Si system where the film is under compressive stress. Taking the misfit strain into consideration, three cases can be considered in epitaxial film growth: large (>10%), intermediate (2% < x < 10%) and small (<2%).

1. In the case of large planar misfit strains (≥10%), the critical thickness is less than a monolayer and therefore, the film grows fully relaxed at growth temperatures with very little residual strain as all of the strains (lattice misfit, thermal misfit and defects) are additive. With subsequent cooling, only thermal and defect strains remain, and the cooling rate plays a critical role in trapping defects associated strains and influencing the dislocation nucleation and propagation.

2. In the case when the planar misfit strains are intermediate, the thickness of the film before dislocations form is several monolayers. Dislocations determine the residual strains, however, if the lattice misfit strains are not relaxed, they will dominate the total residual strain (the contribution from thermal strains is in the order of 0.1–0.2%).

3. When the interplanar misfit strain is small, the pseudomorphic growth can occur, where the film adopts the interplanar spacing of the substrate. As the thickness of the pseudomorphic layer increases, strain-free energy accumulates and, above the critical thickness, it forbids uniform growth and a novel equilibrium structure can sometimes form. In this case, the strain-free energy competes with the chemical free-energy of the metastable pseudomorphic phase to limit the critical thickness [11]. This novel phase can have no residual strain, with misfit dislocations serving as a strain buffer between the pseudomorphic transition layer and a newly formed phase. In this case, the critical thickness is within the order of magnitude of the interplanar spacing due to the large misfit between the new phase and the pseudomorphic transition layer. These pseudomorphic structures have been reported in $TiO_2/Ti_2O_3/Al_2O_3$ and $VO_2/V_2O_3/Al_2O_3$ heterostructures, which follow the planar matching domain epitaxy paradigm [14].

2. Thermal misfit considerations

For a high-quality, epitaxial thin-film growth, the deposition techniques generally required high temperatures and strictly controlled environment [15, 16]. Little consideration was given to parameters such as dislocation formation energy, diffusion and bonding energy. Laws of thermodynamics put limits on what growth conditions are possible, and with high-temperature deposition, the overall process preserves the equilibrium. Thermal misfit strain arises as a result of different coefficients of thermal expansion between the substrate and the film. Thermal misfit strain (ϵ_T) in the film plane is given by

$$\epsilon_T = (\alpha_{Substrate} - \alpha_{film})\Delta T \quad (3)$$

For example, if the expansion coefficient of the film is larger than that of the substrate, the film will have tensile stress upon cooldown and the substrate will experience compressive stress subsequently.

In case of strongly bonded oxide and nitrides, where dislocation formation and propagation steps are difficult, the thermal misfit stress gets more prominent. With increasing the thickness of the film, the strain in the film accumulates and causes cracking and delamination.

3. Thermal processing of thin film heterostructures

As discussed earlier in this chapter, the quality of thin film growth is mainly influenced by lattice misfit and thermal coefficient of expansion coefficient. Considering both factors is the first step toward creating a high-quality, epitaxial interface. Combining the results from literature, three distinct growth categories were proposed by Rasic and Narayan: (i) non-epitaxial samples, (ii) epitaxial films grown on small misfit substrates (LME) and (iii) epitaxial films grown on large misfit substrates (DME) [17] In the following, a summary of the critical considerations for epitaxy across the misfit scale in the presence of lattice/planar misfit, thermal and defect strains is presented. In order to establish each category, a set of three epitaxial (110) films of lanthanum strontium manganese oxide (LSMO) grown with orders of magnitude different oxygen partial pressure conditions were subjected to annealing experiments [17]. The findings indicated that annealing above growth temperature (~900°C) created an irreversible strain relaxation in the films which degraded the magnetization saturation of LSMO films. The hypothesis was further supported by an in-situ XRD experiment that showed a near-linear increase in interplanar spacing (d) until ~690°C for MgO and ~520°C for LSMO layer. Additional increase in the temperature indicated a decrease in the unit cell size of the film and the buffer layer, suggesting that the samples were subjected to both irreversible defect nucleation and recombination, and reduction of oxygen. Partial recovery of magnetic properties was seen with samples being subsequently annealed in pure (99.99999%) oxygen at 700°C for approximately 6 hours. This is assumed to be due to the improved stoichiometry, grain growth and defect annihilation. Nevertheless, XRD showed that characteristic Bragg peaks of the film never returned to the starting value, suggesting that the irreversible process occurred with the high-temperature (900°C) air annealing. Similar thermal experiments in high-vacuum instead of atmosphere were conducted and results indicated a complete collapse of the perovskite crystal structure in the LSMO films suggestive of the significant reduction. Lastly, lower temperature (~500°C) oxidation experiment was conducted (**Figure 3**) on the as-grown sample that resulted in no observable change in the unit cell size, suggesting that the strain state of the film remained constant [17].

From literature, substantial research has been reported on various annealing treatments of LSMO thin films [18–27]. However, correlation between the measured physical properties and structural changes has been elusive. Therefore, It was proposed that the role of annealing treatments on LSMO films can be categorized into three (3) discrete groups: (i) non-epitaxial samples (e.g. nano-crystalline and polycrystalline thin films, and bulk samples), (ii) small lattice misfit epitaxial

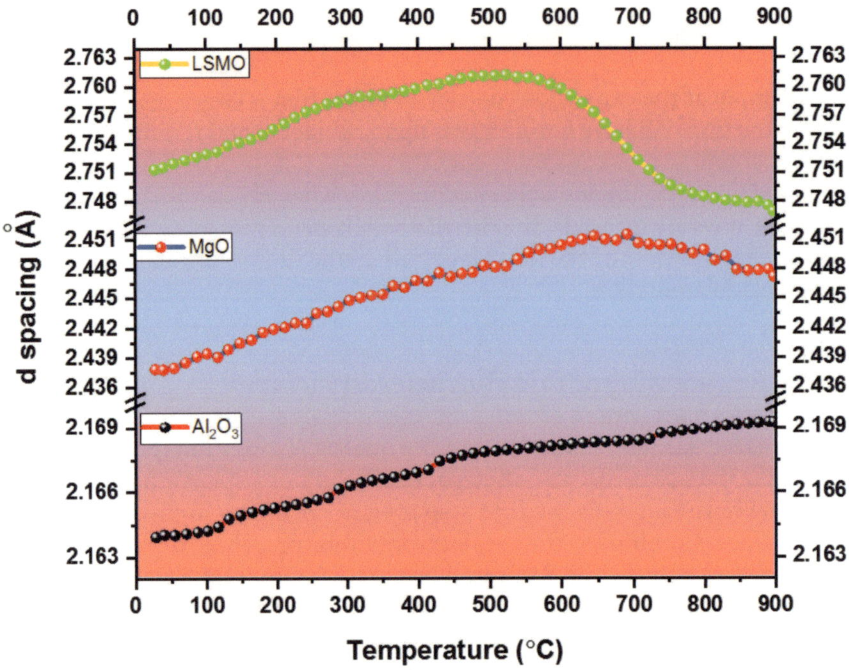

Figure 3.
In-situ X-ray diffraction scan (XRD) showing the change in the interplanar spacing (d) of the (110) LSMO, (111) MgO and (0006) Al_2O_3 Bragg peaks with respect to changing temperature. At ~520°C (LSMO) and ~690°C (MgO) the drop in d spacing is observed, indicating the strain relaxation and oxygen reduction [17].

heterostructures (e.g. STO/LSMO structure) grown by the traditional lattice-matching-epitaxy (LME), and (iii) epitaxial heterostructures grown by the principle of domain matching epitaxy (DME) with a large (>7%) lattice misfit between the film and substrate.

i. In the first (i) category (non-epitaxial samples), high-temperature annealing expectedly increases the grain size, causes defect recombination and annihilation that improves physical properties such as magnetization saturation. Strain relaxation mechanism by dislocations is not present as there is no significant epitaxial relationship between the substrate and the film causing grain growth to be the dominant effect during heat treatments [18, 28].

ii. In small lattice misfit (<7%) substrate/film heterostructures grown by LME (e.g. $SrTiO_3$, $LaAlO_3$ and $BaTiO_3$), the unrelaxed residual lattice misfit strain accumulates with increasing film thickness until the critical thickness where dislocation nucleation and formation starts relaxing the structure [11]. For example, the critical thickness is ~55 nm in the STO/LSMO system [29]. In this example, the film thickness becomes an additional parameter.

(iia) When films are grown with small lattice misfit below the critical thickness value, the total strain that the structure experiences is accumulating and adding to the residual thermal strain.

(iib) When the same heterostructure is grown above the critical thickness, the heterostructure behaves similarly to the large lattice misfit samples (iii), meaning that the residual lattice misfit strain is marginal and the main contribution to the strain is from the mismatch of expansion coefficients between the substrate and the film.

In (iia) and (iib), the subsequent thermal annealing below the growth temperature does not result in a noticeable strain relaxation. Consequently, LSMO films annealed in either oxygen [22] or air [20] did not show significant improvements in magnetic properties. In these cases, the thermal expansion coefficients of the STO ($\sim 1.11 \times 10^{-5}$ K^{-1}) [30] and LSMO ($\sim 1.16 \times 10^{-5}$ K^{-1}) [31] are comparable, resulting in a small change of the overall strain state with temperature. This means that for the small lattice misfit films grown below the critical thickness (iia), the behavior is similar to that of large lattice misfit systems annealed below the growth temperatures (iiia) and the non-epitaxial samples (i). Furthermore, for the small lattice misfit heterostructures grown above the critical thickness (iib), the film will behave similarly to the large lattice misfit epitaxial films that were annealed above the growth temperature (iiib): both systems will relax via dislocation formation [32]. Following enhancement of the physical properties by thermal treatments is due to grain growth and decrease of oxygen vacancies in the film.

iii. The epitaxial thin-film growth dynamic changes significantly with a large lattice misfit systems (>7%) and is explained by domain matching epitaxy paradigm (DME), which predicts strain relaxation via dislocation formation at the substrate/film interface. Firstly, an equilibrium number of dislocations and point defects is present to cancel the lattice misfit strains at the growth temperature. After the deposition of the film, the lattice misfit plays no role in the relaxation mechanism. Instead, the thermal expansion mismatch dictates the residual strain in the film. Hence, the behavior of the film that is undergoing thermal annealing is controlled by the temperature at which the process is occurring and therefore, two independent behaviors are reported.

(iiia) In the case where the sample is subjected to thermal annealing below the original growth temperature, there is no change in the strain state of the sample as all the dislocations remain at the interface and improvements in the physical properties and structure are due to grain growth and point defect annihilation. This behavior is similar to samples in category (i) in the entire temperature range, and category (ii) below the growth temperature.

(iiib) Finally, when annealing the large lattice misfit heterostructure above the original growth temperature, the strain state of the film is "recalculated". In other words, additional formation of dislocation occurs that changes the strain landscape in the film and at the interface. In this case, additional consideration needs to be taken regarding the pressure and composition of the gas in the environment. In the experiments by Rasic and Narayan, the formation of dislocations and film reduction were mutually inclusive at high-temperature thermal processing.

4. Room-temperature epitaxial thin film growth

Reducing the thermal budget of epitaxial thin film growth has been one of the biggest challenges for the electronics industry. By using the pulsed laser deposition technique and starving the system of thermal energy, epitaxial growth is made possible at room temperatures [8]. The ability to use a lower energy budget for epitaxial thin film growth is not only interesting from a cost saving perspective, but also from the diffusion considerations. The semiconductor industry is interested in finding a solution to prevent the electron tunneling phenomena that occurs in a doped silicon once the feature size becomes smaller than 5 nm. Lowering the temperature at which the devices are made will decrease the interfacial diffusion between the functional layers and allow for better performance. The temperature dictates the kinetics and energetics of film growth. Therefore, as a general rule, reducing the thermal budget increases the defect concentration and destroys the epitaxy of the thin films [33–35]. On the other hand, it also decreases the overall cost of fabrication significantly. For these reasons, room temperature growth of high-quality epitaxial thin films that possess minimal viable product characteristics would revolutionize the industry.

In this part of the chapter, a successful room temperature, epitaxial growth of titanium nitride on c-cut (0001) sapphire is reported. The growth is governed by the principle of domain matching epitaxy (DME) where the lattice misfit is ~8.46%. DME theoretical framework was used in this paper to explain the experimental growth. Films were grown at 650°C, 450°C and room temperature (RT). Higher residual out-of-plane strain was observed in room temperature grown films due to the incomplete lattice relaxation.

Scanning transmission electron microscopy (STEM) showed periodic dislocation formation at the film-substrate interface (**Figure 4**), while electron energy loss spectroscopy (EELS) provided insight into interface interdiffusion phenomena at high temperature. An atomically sharp substrate/film interface was observed at room temperature film. High-quality Raman spectra were acquired, confirming that higher nitrogen vacancy concentrations are present with the decreasing temperature of deposition. Further evidence by low-temperature flattening was observed in resistivity vs. temperature measurements, showing that the RT grown

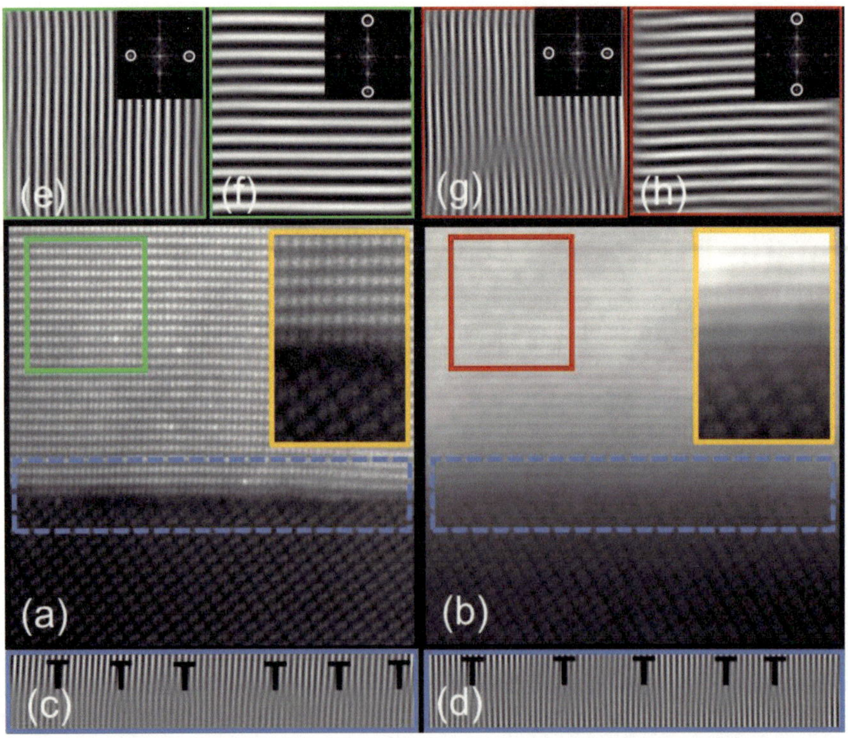

Figure 4.
STEM HAADF images of TiN films grown at (a) high-temperature and (b) room-temperature conditions. The inset images show the interface between TiN and the substrate. IFFT along (110) from the TiN/Al_2O_3 interface region (marked in (a) and (b)) in (c) HT and (d) RT growth [8].

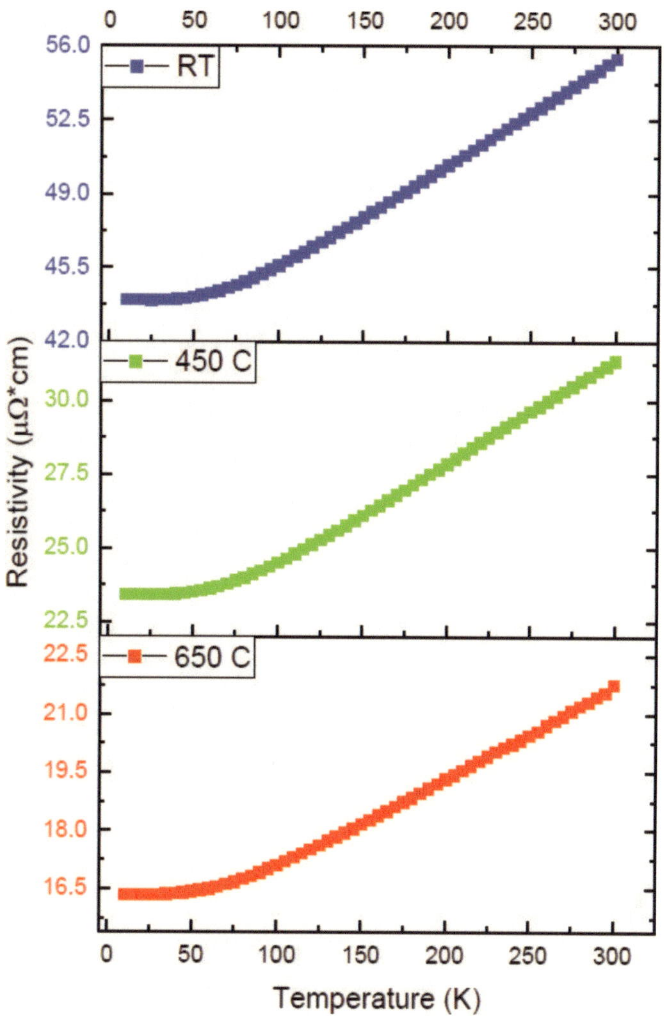

Figure 5.
The resistance vs. temperature measurements of the room-temperature, 450 and 650°C grown TiN samples. The flattening of the RvsT curves below ~35 K is due to defects becoming the dominant scattering mechanism for charge carriers [8].

film displays stronger defect scattering than HT grown films (**Figure 5**). Finally, the resistivity of room-temperature grown film was ~55 μΩ cm as compared to ~22 μΩ cm for 650°C grown films.

Resistivity of TiN films grown at different temperatures is shown in **Figure 5**. Resistivity of the titanium nitride film grown at high temperature was ~22 μΩ cm while room temperature sample demonstrated resistivity of ~55 μΩ cm, indicating comparable values and possibly opening the door for room-temperature epitaxial thin film growth for advanced microelectronic devices that require low energy budget, which will become more important with smaller gate sizes and inter-diffusion between interfaces becomes more prominent.

5. Conclusions

This chapter focused on furthering the current knowledge of thin-film epitaxy; extensive investigations were performed on the importance of deposition temperature and post-deposition annealing while also considering the initial lattice misfit between the substrate and the film.

Post-deposition thermal processing was discussed in the case of the LSMO/MgO/Al_2O_3 heterostructures. It was observed that the behavior of thin films can be categorized into three distinct groups: (i) non-epitaxial samples (e.g. nano-crystalline and polycrystalline films, and bulk samples), (ii) epitaxial films with a small lattice misfit, and (iii) epitaxial films grown with a large lattice misfit between the substrate and the film. This work ties years of thin film annealing research together and provides the underlying theory that translates into all thin film heterostructures. Furthermore, room temperature epitaxial growth was successfully demonstrated on titanium nitride (TiN) on a sapphire substrate. This growth was rationalized by domain matching epitaxy paradigm (DME) and it was made possible at a low temperature due to highly non-equilibrium, pulsed laser deposition growth technique. The room temperature epitaxy demonstrated here transcends from TiN system and can be considered in various heterostructures. Its potential impact in the thin film industry is tremendous. Further research should be on expanding the number of material/substrate possibilities that would allow low-temperature epitaxy creates new directions in all categories of thin film science [8]. Al_2O_3 substrate was established to be the easiest substrate for its low surface reactivity. For example, silicon is highly reactive in atmosphere and will form a silicon dioxide that will not allow epitaxial growth for most materials below 500°C. However, with high-vacuum and targeted cleaning process, silicon and other industrially attractive alternatives can be considered. In the case of sapphire substrates, materials such as zinc oxide (ZnO) [36–38], bismuth ferrite (BFO) [39, 40], barium titanate (BTO) [41], and others [42–47] can be considered.

Acknowledgements

This research was supported by ARO Grant No. W911NF-16-2-0038. The authors are grateful to the Fan Family Foundation Distinguished Chair Endowment for support of J.N. This research was performed in part at the Analytical Instrumentation Facility (AIF) at North Carolina State University, which is supported by the State of North Carolina and the National Science Foundation (Award Number ECCS-1542015). The authors acknowledge the use of the SQUID and PPMS facility in the Department of Materials Science and Engineering at North Carolina State University. The authors would like to thank Dr. John Prater (ARO) for discussing this work and advising.

Conflict of interest

The authors declare no conflict of interest.

DOI: http://dx.doi.org/10.5772/intechopen.82745

Author details

Daniel Rasic* and Jagdish Narayan
North Carolina State University, Raleigh, USA

*Address all correspondence to: drasic@ncsu.edu

References

[1] Moore GE. Cramming more components onto integrated circuits. Proceedings of the IEEE [Internet]. 1998 Jan;**86**(1):82-85. Available from: papers2://publication/uuid/E9E1424B-C5CF-4BE3-866F-4485D525DAB5

[2] Nawrocki W. Physical limits for scaling of integrated circuits. Journal of Physics Conference Series. 1 Nov 2010;**248**(1):012059. Available from: http://stacks.iop.org/1742-6596/248/i=1/a=012059?key=crossref.5b8b79a2026a676da56190726ceee52e

[3] Nawrocki W. Physical Limits for Scaling of Electronic Devices in Integrated Circuits. Dordrecht: Springer; 2011. pp. 79-86. Available from: http://www.springerlink.com/index/10.1007/978-94-007-0044-4_6

[4] Singh RK, Narayan J. Pulsed-laser evaporation technique for deposition of thin films: Physics and theoretical model. Physical Review B. 1990;**41**(13):8843-8859

[5] Lowndes DH, Geohegan DB, Puretzky AA, Norton DP, Rouleau CM. Synthesis of novel thin-film materials by pulsed laser deposition. Science (80-). 1996;**273**(5277):898-903. Available from: http://www.ncbi.nlm.nih.gov/pubmed/8688065

[6] Eason R. Pulsed Laser Deposition of Thin Films: Applications-led Growth of Functional Materials. Wiley-Interscience; 2007. 682 p. Available from: https://books.google.com/books?id=0jIO3cwkiOIC&dq=PLD+pulsed+laser+deposition&lr=&source=gbs_navlinks_s

[7] Venkatesan T, Wu X, Am AI, Chang CC, Hegde MS, Dutta B. Laser processing of high-superconducting thin films. IEEE Journal of Quantum Electronics. 1989;**25**(11):2388-2393. Available from: http://ieeexplore.ieee.org/document/42070/

[8] Rasic D, Sachan R, Chisholm MF, Prater J, Narayan J. Room temperature growth of epitaxial titanium nitride films by pulsed laser deposition. Crystal Growth & Design. 2017;**17**(12):6634-6640. Available from: http://pubs.acs.org/doi/abs/10.1021/acs.cgd.7b01278

[9] Lal K, Meikap AK, Chattopadhyay SK, Chatterjee SK, Ghosh M, Baba K, et al. Electrical resistivity of titanium nitride thin films prepared by ion beam-assisted deposition. Physica B: Condensed Matter. 2001;**307**(1-4):150-157. Available from: http://linkinghub.elsevier.com/retrieve/pii/S0921452601006263

[10] Chowdhury R, Vispute RD, Jagannadham K, Narayan J. Characteristics of titanium nitride films grown by pulsed laser deposition. Journal of Materials Research. 1996;**11**(001):1458-1469

[11] Narayan J. Recent progress in thin film epitaxy across the misfit scale (2011 Acta Gold Medal Paper). Acta Materialia. 2013;**61**(8):2703-2724. Available from. DOI: 10.1016/j.actamat.2012.09.070

[12] Rasic D, Sachan R, Temizer NK, Prater J, Narayan J. Oxygen effect on the properties of epitaxial (110) $La_{0.7}Sr_{0.3}MnO_3$ by defect engineering. ACS Applied Materials & Interfaces. 2018;**10**(24):21001-21008. DOI: 10.1021/acsami.8b05929

[13] Ichimuraa M, Narayan J. Atomistic study of dislocation nucleation in Ge/(001)Si heterostructuses. Philosophical Magazine A. 1995;**72**(2):281-295. DOI: 10.1080/01418619508239925

[14] Domain epitaxy for thin film growth. 2005. Available from: https://patents.google.com/patent/US6955985B2/en

[15] Balevičius S, Tornau EE, Žurauskienė N, Stankevič V, Šimkevičius Č, Tolvaišienė S, et al. Magnetoresistance anisotropy of ultrathin epitaxial $La_{0.83}Sr_{0.17}MnO_3$ films. Journal of Applied Physics. 2017;**122**(101). DOI: 10.1063/1.4998232

[16] Martin LW, Schlom DG. Advanced synthesis techniques and routes to new single-phase multiferroics. Current Opinion in Solid State & Materials Science. 2012;**16**(5):199-215

[17] Rasic D, Sachan R, Prater J, Narayan J. Structure-property correlations in thermally processed epitaxial LSMO films. Acta Mater [Internet]. 15 Jan 2019;**163**:189-198. [cited 2018 Oct 22]. Available from: https://www.sciencedirect.com/science/article/pii/S1359645418308152?dgcid=author

[18] Thoma P, Monecke M, Buja OM, Solonenko D, Dudric R, Ciubotariu OT, et al. Polycrystalline $La_{1-x}Sr_xMnO_3$ films on silicon: Influence of post-deposition annealing on structural, (magneto-) optical, and (magneto-)electrical properties. Applied Surface Science. 2018;**427**:533-540. DOI: 10.1016/j.apsusc.2017.08.046

[19] Shinde KP, Pawar SS, Pawar SH. Influence of annealing temperature on morphological and magnetic properties of $La_{0.9}Sr_{0.1}MnO_3$. Applied Surface Science. 2011;**257**(23):9996-9999. DOI: 10.1016/j.apsusc.2011.06.126

[20] Ichinose T, Naganuma H, Miyazaki T, Oogane M, Ando Y, Ueno T, et al. Effect of annealing on curie temperature and phase transition in $La_{0.55}Sr_{0.08}Mn_{0.37}O_3$ epitaxial films grown on $SrTiO_3$ (100) substrates by reactive radio frequency magnetron sputtering. Materials Characterization. 2016;**118**:37-43. DOI: 10.1016/j.matchar.2016.05.002

[21] Li T, Wang B, Dai H, Du Y, Yan H, Liu Y. Annealing effect on the structural and magnetic properties of $La_{0.7}Sr_{0.3}MnO_3$ films. Journal of Applied Physics. 2005;**98**(12):2003-2006

[22] Du YS, Wang B, Li T, Yu DB, Yan H. Effects of annealing procedures on the structural and magnetic properties of epitaxial $La_{0.7}Sr_{0.3}MnO_3$ films. Journal of Magnetism and Magnetic Materials. 2006;**297**(2):88-92

[23] Xie H, Huang H, Cao N, Zhou C, Niu D, Gao Y. Effects of annealing on structure and composition of LSMO thin films. Physica B: Condensed Matter. 2015;**477**:14-19. DOI: 10.1016/j.physb.2015.07.032

[24] Wu W, Wong KH, Li X-G, Choy CL, Zhang YH. Effect of annealing in reduced oxygen pressure on the electrical transport properties of epitaxial thin film and bulk $(La_{1-x}Nd_x)_{0.7}Sr_{0.3}MnO_3$. Journal of Applied Physics. 2000;**87**(6):3006 Available from: http://link.aip.org/link/JAPIAU/v87/i6/p3006/s1&Agg=doi

[25] Nakamura K, Liu X, Hatano T, Jiao Z, Shang K, Ishii A. Mn and Mg interdiffusion and magnetotransport properties of $La_{0.7}Sr_{0.3}MnO_3$ films on MgO (100) substrate. Japanese Journal of Applied Physics. 2000;**39**(4R):1721 Available from: http://stacks.iop.org/1347-4065/39/i=4R/a=1721

[26] Sirena M, Haberkorn N, Granada M, Steren LB, Guimpel J. Oxygen and disorder effect in the magnetic properties of manganite films. Journal of Magnetism and Magnetic Materials. 2004;**272-276**:1171-1173

[27] Sahu DR, Mishra DK, Huang JL, Roul BK. Annealing effect on the properties of $La_{0.7}Sr_{0.3}MnO_3$ thin film grown on Si substrates by DC sputtering. Physica B: Condensed Matter. 2007;**396**(1-2):75-80

[28] Zi Z, Tang X, Zhang M, Dai J, Sun Y. Modified magnetic properties

of $La_{0.7}Sr_{0.3}MnO_3$ film by magnetic-field annealing. Materials Letters. 2016;**169**:200-202

[29] Pradhan AK, Hunter D, Williams T, Lasley-Hunter B, Bah R, Mustafa H, et al. Magnetic properties of $La_{0.6}Sr_{0.4}Mn_{O3}$ thin films on $SrTiO_3$ and buffered Si substrates with varying thickness. Journal of Applied Physics. 2008;**103**(2):023914. DOI: 10.1063/1.2833388

[30] de Ligny D, Richet P. High-temperature heat capacity and thermal expansion of $SrTiO_3$ and $SrZrO_3$ perovskites. Physical Review B. 1996;**53**(6):3013-3022

[31] Hammouche A, Siebert E, Hammou A. Crystallographic, thermal and electrochemical properties of the system $La_{1-x}Sr_xMnO_3$ for high temperature solid electrolyte fuel cells. Materials Research Bulletin. 1989;**24**(3):367-380

[32] Seo SH, Kang HC, Jang HW, Noh DY. Effects of oxygen incorporation in tensile $La_{0.84}Sr_{0.16}MnO_{3-\delta}$ thin films during ex situ annealing. Physical Review B: Condensed Matter and Materials Physics. 2005;**71**(1):2004-2006

[33] Ranno L, Llobet A, Hunt M, Pierre J. Influence of substrate temperature on magnetotransport properties of thin films of $La_{0.7}Sr_{0.3}MnO_3$. Applied Surface Science. 1999;**138-139**:228-232 Available from: http://www.sciencedirect.com/science/article/pii/S0169433298004000

[34] Chen W-C, Wu S-T. Epitaxial growth of TiN on Al_2O_3 at cryogenic temperature. Applied Physics. 2003;**42**(2):192-193

[35] Lackner JM. Industrially-scaled room-temperature pulsed laser deposition of Ti-TiN multilayer coatings. Journal of Physics Conference Series. 2007;**59**:16-21 Available from: http://stacks.iop.org/1742-6596/59/i=1/a=004?key=crossref.880096637b9988533e66a81588420afc

[36] Sun XW, Kwok HS. Optical properties of epitaxially grown zinc oxide films on sapphire by pulsed laser deposition. Journal of Applied Physics. 1999;**86**(1):408-411. DOI: 10.1063/1.370744

[37] Kaidashev EM, Lorenz M, Von Wenckstern H, Rahm A, Semmelhack HC, Han KH, et al. High electron mobility of epitaxial ZnO thin films on c-plane sapphire grown by multistep pulsed-laser deposition. Applied Physics Letters. 2003;**82**(22):3901-3903. DOI: 10.1063/1.1578694

[38] Srikant V, Sergo V, Clarke DR. Epitaxial aluminum-doped zinc oxide thin films on sapphire: I. Effect of substrate orientation. Journal of the American Ceramic Society. 1995;**78**(7):1931-1934. DOI: 10.1111/j.1151-2916.1995.tb08912.x

[39] Avjyan KE, Buniatyan VV, Dashtoyan HR. Optical properties of nanoscale $BiFeO_3/BaTiO_3/Ni_{0.5}Zn_{0.5}Fe_2O_4$ composite films obtained by the pulsed-laser deposition method. Journal of Contemporary Physics (Armenian Academy of Sciences). 2013;**48**(3):134-137. DOI: 10.3103/S1068337213030079

[40] Zhao YJ, Yin ZG, Zhang XW, Fu Z, Sun BJ, Wang JX, et al. Heteroepitaxy of tetragonal $BiFeO_3$ on hexagonal sapphire(0001). ACS Applied Materials & Interfaces. 2014;**6**(4):2639-2646. DOI: 10.1021/am405115y

[41] Burch MJ, Li J, Harris DT, Maria JP, Dickey EC. Mechanisms for microstructure enhancement in flux-assisted growth of barium titanate on sapphire. Journal of Materials Research. 2014;**29**(7):843-848 Available from: http://www.journals.cambridge.org/abstract_S0884291414000594

[42] Ueda K, Hase T, Yanagi H, Kawazoe H, Hosono H, Ohta H, et al. Epitaxial growth of transparent p-type

conducting $CuGaO_2$ thin films on sapphire (001) substrates by pulsed laser deposition. Journal of Applied Physics. 2001;**89**(3):1790. DOI: 10.1063/1.1337587

[43] Dominguez JE, Pan XQ, Fu L, Van Rompay PA, Zhang Z, Nees JA, et al. Epitaxial SnO_2 thin films grown on ($\bar{1}012$) sapphire by femtosecond pulsed laser deposition. Journal of Applied Physics. 2002;**91**(3):1060-1065. DOI: 10.1063/1.1426245

[44] Dominguez JE, Fu L, Pan XQ. Epitaxial nanocrystalline tin dioxide thin films grown on (0001) sapphire by femtosecond pulsed laser deposition. Applied Physics Letters. 2001;**79**(5):614-616. DOI: 10.1063/1.1386406

[45] Marsh AM, Harkness SD, Qian F, Singh RK. Pulsed laser deposition of high quality $LiNbO_3$ films on sapphire substrates. Applied Physics Letters. 1993;**62**(9):952-954. DOI: 10.1063/1.108530

[46] Vispute RD, Talyansky V, Sharma RP, Choopun S, Downes M, Venkatesan T, et al. Growth of epitaxial GaN films by pulsed laser deposition. Applied Physics Letters. 1998;**71**(1):102. DOI: 10.1063/1.119441

[47] Samarasekara P, Rani R, Cadieu FJ, Shaheen SA. Variable texture $NiOFe_2O_3$ ferrite films prepared by pulsed laser deposition. Journal of Applied Physics. 1998;**79**(8):5425. DOI: 10.1063/1.362327

Chapter 6

Conventional and Unconventional Crystallization Mechanisms

Kamila Ferreira Chaves, Thaís Jordânia Silva,
Maria Aliciane Fontenele Domingues, Daniel Barrera-Arellano
and Ana Paula Badan Ribeiro

Abstract

Understanding the crystallization behavior of fats and oils is essential to ensure certain desirable characteristics in a given industrial application. In recent years, some advances in the structuring of lipid phases have enabled a direct influence on the food properties. The structuring mechanisms of lipid bases can be classified as either conventional or unconventional. Conventional crystallization mechanisms consist of nucleation, growth, and maturation of the crystals, thus resulting in a crystalline lattice. Co-crystallization or seeding agents and emerging technologies such as ultrasound can be used to aid in crystallization and improve the physical properties of fats and oils. Unconventional mechanisms bring organogel technology as a trend, which consists in the use of self-assembly agents to entrap the liquid oil, resulting in a structured gel network. In this chapter, the formation process of crystalline networks and gel networks will be presented in stages, highlighting the main differences related to the mechanisms of formation and stabilization of both types of networks.

Keywords: conventional crystallization, unconventional crystallization, crystallization mechanisms, crystalline networks, gel networks, seeding agents, emerging technologies

1. Introduction

Edible oils and fats are essential nutrients in the diet as they provide essential fatty acids and energy. In addition, they have important functions in the production of processed foods. These ingredients are directly related to texture, stability, aeration, and nutritional features of the products [1, 2]. Some fats used in food are known by the term shortening, also known as plastic fats. Selection of the best shortening is based on formulations (product to which it is intended), functions, and process conditions. The production of shortening is related to the types of fats and oils used and the way in which they crystallize and form solid networks. This is because solid crystalline fat networks are responsible for providing the expected physical macroscopic functions of the various types of shortening systems, due to the way they melt and to the hardness, consistency, and complacency of the network [3].

Oils and fats are one of the three food macronutrients, represented by triacylglycerol (TAG) molecules, formed by an ester bond between glycerol and three fatty acids. TAG crystals are known to show polymorphism, which is the ability of a substance to have more than one structure or crystalline form. As the physical properties of fat crystals, such as melting point and rheology, are important for food texture, polymorphic control of fat crystals is crucial in the food industry, and it is thus necessary to understand the crystallization mechanisms of TAGs.

Crystallization consists of an arrangement of the system as a result of a driven force, characterized by the total or partial restriction of movement caused by physical or chemical bonds between TAG molecules. Due to the complexity of this system, TAGs can be compacted in different ways. This crystallization has implications in the industrial processing of foods, since these depend on fat crystals in products like margarines, chocolate, spreads, baked goods, etc. [4, 5].

2. Lipids

Lipids are defined as compounds soluble in organic solvents and insoluble in water. These compounds are represented by oils and fats. Chemically, vegetable oils and fats are mainly composed of TAGs and may contain small amounts of diacylglycerols, monoacylglycerols, free fatty acids, phospholipids, waxes, and unsaponifiable constituents [6].

Glycerol has three positions in its molecule (*sn*-1, *sn*-2, and *sn*-3) in which the fatty acids are esterified to form the TAG. The features of these fatty acids influence the physical properties of oils and fats: (i) in relation to unsaturation and saturation: saturated fatty acids have a higher melting point compared to unsaturated ones, since unsaturation impairs the packing of molecules; (ii) in relation to the geometric isomerism: it refers to the possible arrangements of the hydrogen atoms around the double bond, presenting the *cis* or *trans* form. *Cis* bonds reduce the melting point due to folding of the molecule, impairing packing. Conversely, fatty acids in a *trans* configuration have a melting point close to that of saturated fatty acids, due to torsion in the molecule, making it linear; and (iii) in relation to chain size: the longer the chain, the higher the melting point. Physically, oils are liquid at room temperature, while fats are solid, semi-solid, and plastic at room temperature. Therefore, the composition in TAGs determines the physical and functional properties of oils and fats, affecting the structure, stability, taste, and the sensorial and visual characteristics of foods [5]

3. Crystallization mechanisms

3.1 Conventional crystallization

Crystallization, melting, and recrystallization of lipids are first order transitions of interest in food systems that affect the shelf life of food and food ingredients. The crystallization properties of food components may be highly complex due to soluble-solvent interactions, mixture of miscible components, and polymorphic transitions. The production of solid fats, spreads, and chocolates is an excellent example of the use of first-order transition behaviors of food systems to manipulate properties in processing, structure formation, and application performance [7]. Understanding the development of structures and the control of the crystallization of food components has evolved over the years, but there is still much to learn.

In lipids, the conventional crystallization process is fundamentally important to understand what happens while processing foods such as chocolates, spreads, bakery products, margarines, ice cream, etc. How can these crystals have such a great influence on the quality of a product? We can propose ice cream as an example.

When consuming an ice cream, we want it to be creamy without crystals leaving a sandy sensation in the mouth. Another classic example is chocolate, as a properly crystallized cocoa butter of good quality, which provides a shiny and smooth chocolate that melts homogeneously [8].

Macroscopic properties of lipids in foods are particularly influenced by the microstructure of fats. Therefore, it is important to consider the effects of the microstructure for subsequent analysis of macroscopic properties. The formation and growth rate of the crystals and the polymorphic transformations are important to determine the function and application of oils and fats [4]. Therefore, understanding crystallization is fundamental to obtain products with adequate functions, texture and sensory qualities and that are physically stable.

The following sections will address the mechanisms that occur in vegetable fats during their crystallization. We will also understand that some techniques can induce or delay this crystallization in different ways, in order to achieve the industrially desired result.

3.1.1 Definitions and mechanisms

Before discussing the conventional crystallization mechanism of vegetable fats, we need to define some concepts that are fundamental for a good understanding of the subject. We define conventional crystallization as an arrangement of the system characterized by total or partial restriction of movement, caused by the physical or chemical bonds between the molecules of TAGs. These bonds will lead to what we call crystals, which are molecules arranged in fixed patterns. Its high degree of molecular complexity allows the same set of TAGs to be compacted into several different and relatively stable structures [9, 10].

We can say that the crystallization behavior involves four phases: nucleation (formation of a crystalline phase in the liquid phase through the organization of TAGs in a crystalline network, that is, molecules in the liquid state bind to create a stable nucleus), growth (caused by the inclusion of other TAGs in the existing crystalline network), recrystallization (structural and chemical change of the crystals during storage), and polymorphism (which controls the order of packing of TAG molecules in the crystals). Before forming the first crystals, the system, in its liquid state, needs to reach the supersaturated zone, in which there is a driving force for crystallization [11, 12]. From this point, crystallization starts.

3.1.1.1 Nucleation

It starts with the appearance of a driving force to overcome the energy barrier for crystallization. As a result, molecules in the liquid state somehow bind to create a stable nucleus. Nucleation can be classified as (a) primary nucleation (homogeneous or heterogeneous) (**Figure 1**) and (b) secondary nucleation. Homogeneous primary nucleation occurs with the binding of isolated molecular species, which form dimers and trimers, and subsequently continues the accumulation process until a potential nucleus can be formed. That is, it is a molecular arrangement in a crystalline network, without external aid.

Heterogeneous primary nucleation occurs when the local arrangement of molecules to form the nucleus arises from interactions between the solid particle and the supersaturated fluid, by supplying external energy. Secondary nucleation is the formation of a new nucleus in the presence of existing crystals, which may occur if crystalline microscopic elements are separated from an already formed crystalline surface, resulting in fracture of crystals in small stable nuclei [5, 8]. Once a crystal core is formed, it starts growing due to the incorporation of other molecules of

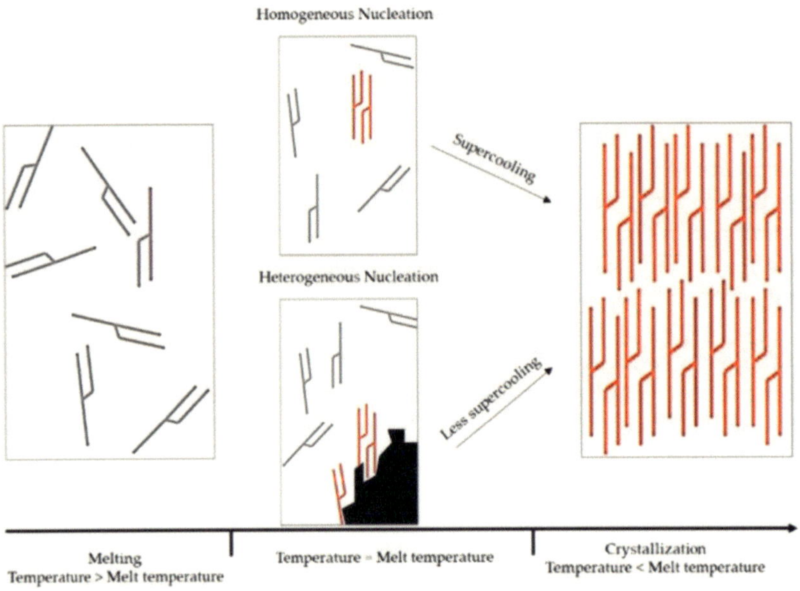

Figure 1.
Nucleation mechanisms: crystal embryo formation in homogeneous nucleation and seeding or co-crystallization in heterogeneous nucleation (adapted from [13]).

the adjacent liquid layer, which is continuously filled by the supersaturated liquid surrounding the crystal [12].

3.1.1.2 Growth

According to the theory, crystalline growth occurs through two features: (1) surface of the growth process and (2) transport process in the face of the growing crystal [14]. The mechanism is based on how the face of the crystal incorporates growth units of the adjacent solution. According to Hartman and Perdok [15], for a three-dimensional crystalline face, the capture of growth units can be classified according to how many binding interactions form between the adsorbed growth unit and the surface of the crystal. If three orientations of surface bonds are involved, the incorporation is then of the kinked face type; if two orientations of surface bonds are involved, the incorporation is of the stepped face type; and if only one surface bond is involved, the incorporation is of the flat face type (**Figure 2**).

Once a crystal nucleus is formed, it starts growing due to the incorporation of other molecules of the adjacent liquid layer that is continuously filled by the supersaturated liquid that surrounds the crystal [12]. When the formed nuclei reach favorable dimensions, these elements become crystallites, and their growth depends on external (supersaturation, solvents, temperature, and impurities) and internal (structure, bonds, and defects) factors. Therefore, the crystal growth rate can vary by several orders of magnitude. Growth occurs through the binding of molecules to a crystalline surface. While the molecules are attached to the surface of a crystal, some molecules are also deactivated. There is a continuous movement of molecules at the surface of the crystal, and the result of these processes determines the growth rate, which is directly proportional to the subcooling and varies inversely to the viscosity system [5, 12]. Unless there is a kinetic constraint, growth continues until the system reaches equilibrium (driving force for crystallization of zero and maximum volume of the phase of the crystals) [8].

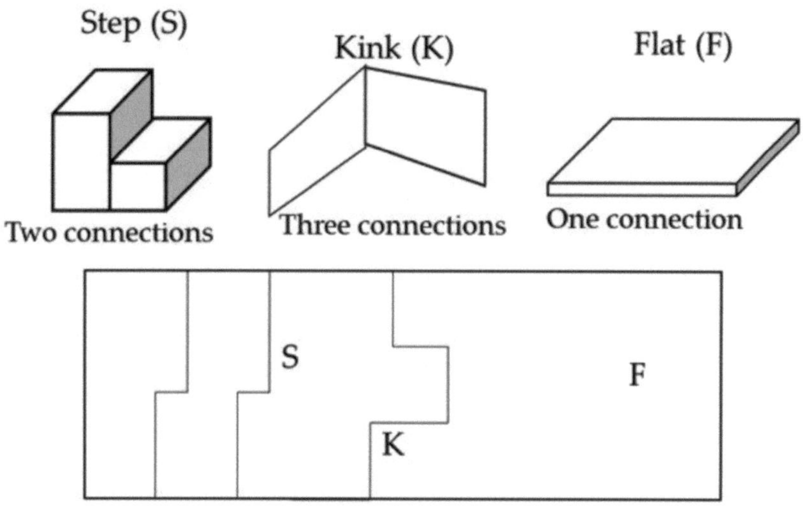

Figure 2.
Classification of surface contacts according to growth theory (adapted from [15]).

3.1.1.3 Recrystallization

The final stage of crystallization in food occurs during storage, when crystals undergo a recrystallization step [8]. This phenomenon can be defined as a change in the size, shape, orientation, and perfection of the crystals [16]. Basically, in this stage, small crystals, which are more soluble, disappear and larger crystals grow. The concern with this step is related to the changes in the products during their storage, such as fat bloom in chocolates and coatings and oil exudation in fat-rich products [12].

3.1.1.4 Polymorphism

When discussing lipid crystallization, we must relate this phenomenon to one of the most important physical properties, which is polymorphism. Polymorphism can be defined in terms of the ability to present different unit cell structures due to several molecular packages [5]. In lipids, there are three specific types of subcells (the smaller periodic structure that exists in the current cellular unit, which is defined as the transverse mode of packing of the aliphatic chains in the TAGs), the polymorphs: α, β′, and β [4]. The α form is metastable, with hexagonal chain packing. The β′ form has intermediate stability and orthorhombic perpendicular packing. Crystals in the β′ form show greater function, since they are smoother and provide good aeration and creaminess properties. The β form has greater stability and triclinic parallel packing. The melting temperature increases with increasing stability ($\alpha \rightarrow \beta' \rightarrow \beta$) due to differences in molecular packing density [5, 17]. The crystal structures also differ in chain length, which illustrates the repetitive sequence of acyl chains in a unit cell along the axis of a long chain. The chain length structure ranges from double (DCL) to triple (TCL), fourth (QCL), or hexa (HCL). In addition, the conformation of the glycerol group also influences the phase behavior of fats. Two typical glycerol conformations, fork type and chair type, have been observed in TAG crystals. The fork-like conformation is favored by the presence of two identical fatty acids at positions 1 and 3 of the glycerol, forming a symmetrical TAG. On the other hand, the chair-type conformation is favored in asymmetric TAGs. Neighboring molecules of TAG crystals are stacked in layers, with the parallel hydrocarbon chains packed together and arranged more or less perpendicular to the parallel planes of the glycerol groups and to the methyl terminal group planes. The distances between these molecules characterize the short spacings and the long spacings, providing the polymorphic properties and melting point of each type of fat. The choice between the fork- or chair-like conformation is dictated by the need to accommodate fatty acids of different chain lengths and the distortions introduced by the double bonds [18, 19] (**Figure 3**).

3.1.2 Crystallization modifiers

The crystallization process can be modified using techniques that change the kinetics of the crystals, such as the use of additives and ultrasound.

3.1.2.1 Additives

Lipophilic additives are used as they influence the volume properties (consistency and texture), promoting or inhibiting crystallization. These compounds provide a modeling effect, i.e., seeding nuclei. For example, a higher melting point additive with chemical and structural similarities to the lipid is used as a seeding nucleus for heterogeneous crystallization [20]. Examples of additives that may promote or inhibit crystallization are minor lipids, such as diacylglycerols, monoacylglycerols, free fatty acids, phospholipids, and sterols. These lipids can interact with TAGs in the growth stage, causing a structural competitive effect or permanent incorporation into the crystalline structure, affecting the crystallization rate, polymorphic forms, and microstructure of the crystals, positively or negatively [21].

Another technique that is being increasingly used is seeding. It consists of the addition of a solid material with crystallization nucleating agent properties. The incorporation of crystallization seeds into liquid fats can promote two effects associated with the control of crystallization by the availability of numerous additional nuclei and/or surfaces for crystal growth. In addition, it may promote specific polymorphic forms [21].

The addition of emulsifiers changes the supersaturation characteristics of the solution and consequently its solubility, and thus, the incorporation of growth units on the surface will also change. At low supersaturations, growth occurs by the incorporation of the flat face type, and, in general, the free energy (ΔG) decreases

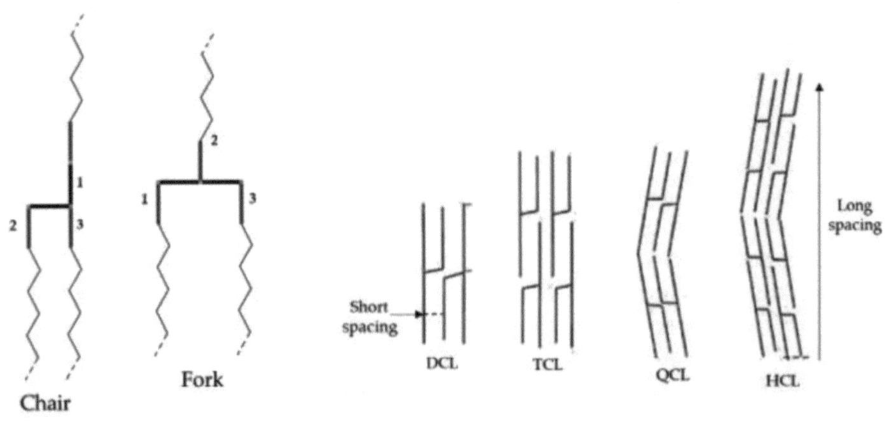

Figure 3.
Conformation of glycerol groups in triacylglycerol crystals and chain length structures of triacylglycerol crystals (adapted from [18]).

according to the sequence kinked < stepped < flat [21]. Thus, the mechanisms of co-crystallization and seeding occur, leading to the template effect of the emulsifiers (**Figure 4**).

3.1.2.2 Ultrasound

This technique has been noted as an excellent alternative to promote crystallization, since the ultrasonic energy is sufficient to promote nucleation and growth by increasing the transfer of heat and mass. Through manipulation of nucleation and growth, ultrasounds can potentially provide improved control for the assembly of crystal structures in foods to control texture or improve separation [8]. It is believed that these ultrasonic effects are superior to conventional agitation to control and stimulate the reaction rate, supersaturation level, nucleation, and crystal growth [22].

Care should be taken in relation to the intensity of ultrasound used, since the crystalline structure is highly dependent on the intensity of the ultrasound. For example, a palm oil where the ultrasound was applied at an intensity of 30 dB had little effect on kinetics but prevented the formation of large spherulitic crystals. When this intensity was increased to 35 dB, it produced smaller, more uniform crystals that tend to clump and fall to the bottom of the cell. A further increase in intensity to 40 dB, below the cavitational threshold, produced a uniform product of very small crystals, suggesting that there are numerous nucleation sites that produced crystals simultaneously [23].

The effects caused by the ultrasound include: (i) faster primary nucleation, which is quite uniform through the sonicated volume; (ii) relatively easy nucleation in materials that are otherwise generally difficult to nucleate; (iii) start of the secondary nucleation; and (iv) production of smaller, purer crystals of more uniform size. In addition, ultrasound can reduce clumping of crystals. The shock

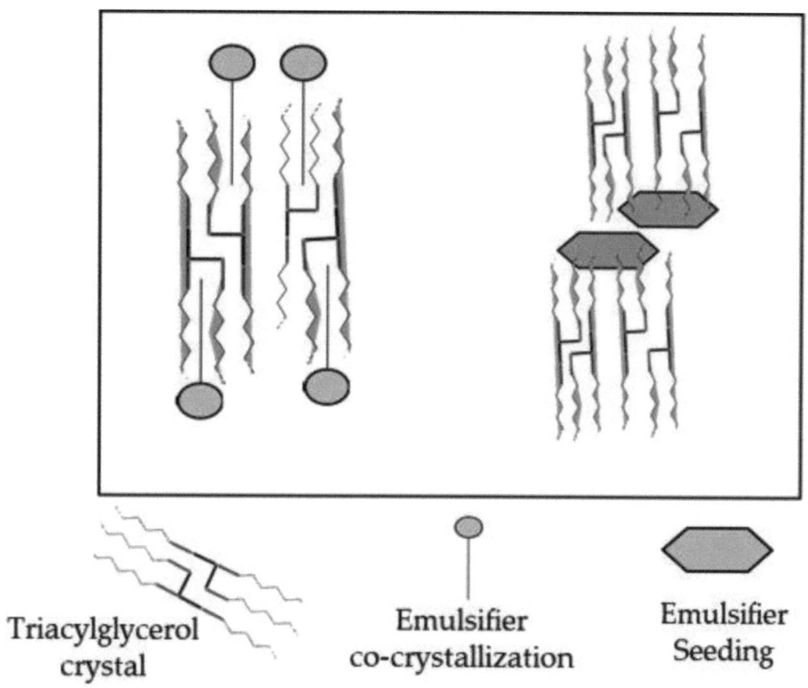

Figure 4.
Template effect of emulsifiers: co-crystallization and seeding.

wave, which is caused by cavitation, decreases the contact between the crystals, preventing clumping, by controlling the population of local nuclei. The induction time is defined as the time elapsed between onset of supersaturation and appearance of crystals and decreases as supersaturation increases. This induction time is drastically reduced by the presence of sonication. This effect is more significant at low absolute supersaturations [24].

3.2 Unconventional crystallization

Studies with organogels led to the development of many applications in the food area, such as chocolate filling, margarines, baked products, such as biscuits and cookies, pastries, spreads, sausages, ice cream, yogurt, unprocessed and processed cheese, mayonnaise, and sauces, among other products [25–28]. Considering this, it is crucial to understand how unconventional crystallization occurs.

Many organogels are more desirable to consumers since they have a better nutritional profile than conventional fats. Therefore, organogel technology seems to be a potential solution for the reduction of saturated fatty acids and possibly the elimination of trans-fatty acids from processed foods [26, 29, 30].

This technology is feasible in comparison to the technologies currently used, since it does not cause any chemical changes in the structure of fatty acids and TAGs and maintains the nutritional features of the oil; in particular, it maintains unsaturated fatty acid contents and natural regiospecific distribution without increasing the AGS content [31].

3.2.1 Definitions and mechanisms

Organogels are a class of materials that can hold large volumes of organic liquids in self-assembled networks and have many practical applications in the pharmaceutical, chemical, and food industries regarding lipid technology [32]. Organogels are formed by structurants of low molecular weight and some organic solvents that are supported, through a three-dimensional network, forming the gel. Since organogels are biodegradable, they can be used in drugs as protein carriers and vaccines for example [33–35].

The formation of these structures through supramolecular assembly is of great interest, since they can form semisolid phases that are produced at low cost [32]. Several researchers have focused their studies on the properties of structurants to form organogels [22, 36–38].

In unconventional structuring, lipid systems composed of unsaturated TAGs, such as liquid or semi-solid vegetable oils, can be structured as gels, forming continuous networks of small molecules that assemble in liquid crystals, micelles, or fibrillar networks formed from aggregates of micelles, developing inverse bilayer structures in the form of rods [39].

Organogels can be structured forming a fibrous 3D network, where the solvent is trapped in the structuring matrix, avoiding the flow of solvent. The network is stabilized by weak interactions between the chains, such as hydrogen bonds, van der Waals forces, and π staking [40–43]. Although it is known that organogels are formed through weak intermolecular interactions between the structurant molecules, which generate three-dimensional networks [44], there is still a lack of fundamental understanding of the type of interactions that are required [32].

The physically driven process of structuring organogels depends on many factors, such as structural adjustment of structurant molecules, solvent effects, concentration, temperature, use of ultrasounds, and shear. When these factors change, the structuring properties are affected due to the precise balance between the interaction of the structurant molecules, the interaction between the solvent and the structurant, and the applied external stimulus [45, 46].

The structurants are insoluble in almost all solvents at room temperature, and the solvent-structurant and structurant-structurant interactions are the main force for the formation of organogels. However, they dissolve after heating and jellify after cooling; thus, the gelation process is an equilibrium between dissolution and precipitation in the solvents [47].

Organogel networks may range from the assembly of surfactants into solution by physical interactions (e.g., micellization, lyotropism, and crystallization) [48], also known as molecular organogels, to the formation of flexible polymer networks (e.g., swelling), known as polymer organogels [49]. The molecular organogels known today are at the interface between complex fluids [50] and solids and, regardless of the nature of the structure, are composed of thermoreversible semi-solid materials [32].

3.2.2 Structuring modifiers

The group of molecular organogels can be subdivided according to the mode of self-organization of the structurant in liquid crystals [51], platelet crystals, elliptic networks [52, 53], and inverted micelles [54, 55].

Several solvent parameters have now been used to correlate with gelation ability [56], such as protic, dipolar aprotic, and apolar aprotic (low polarity), which were too broad to quantify solvent effects [57]. More specifically, solvent parameters are divided into three categories: physical properties, solvatochromic properties, and thermodynamic properties, including dielectric constant, Reichardt ET-30 parameter [58], Kamlet-Taft parameters [59], Hildebrand solubility parameter [60], and Hansen solubility parameters [61, 62].

The structures of the organogel molecules significantly influence the resulting gelation properties. The length of the alkali chain, the position of the chiral units, the substitution groups, and the peptide sequences, in addition to the arrangement or the number of the hydrogen bonds, may influence the properties of organogels [45].

The structures can be formed from amphiphilic compounds that can self-assemble in different microstructures, such as micellar and bilayer phases, which may accumulate in different lyotropic mesophases of larger scales (hexagonal, cubic, or lamellar liquid crystalline phases) at high concentrations [63, 64]. The formation of longer structures provides viscoelastic behavior or a gel-like solution. These viscoelastic materials, created through the assembly of structurants, are described in the literature as supramolecular organogels [65]. The formation mechanism of these structures is based on the arrangement of monomeric units that bind through non-covalent bonds, such as van der Waals interactions, hydrogen bonds, electrostatic interactions, and π-π or τ-stacking [65, 66].

Two different types of packing models were discovered in non-polar and polar solvents. In non-polar solvents, an interdigitated stacking model was preferred in the self-organized 3D micromorphology of the gels, due to inverse strong hydrogen bonds and weak π-π stacking interactions. Conversely, in polar solvents, the structurant is automatically assembled in a 3D nanostructure through a stacking model due to strong π-π stacking interactions and weak hydrogen bonds [67].

In general, there are a limited number of biocompatible components known to structure lipids through the assembly of molecules [46, 68, 69].

The extensive applications of organogels are of great importance to design and study new structurants and to characterize and determine the practical applications for organogels. It is known that some organogels with excellent mechanical and optical properties are more useful for practical applications [45].

Current knowledge on assembly behavior and subsequent gel formation and the effect of salts was applied to induce the formation of networks in hydrophobic environments using an unsaturated fatty acid, oleic acid, and sodium acetate, in sunflower oil. Oleic acid alone did not provide gelation in the oil, but the addition of sodium oleate induced gel formation. In mixtures containing oleic acid, reverse micellar structures are present. The participation of the sodium oleate in the original spherical micelles of oleic acid changes the geometry and assembly behavior of the structurant. At higher concentrations of sodium oleate, lamellar crystal structures are formed. The assembly is theoretically driven largely by hydrophilic interactions between the hydrophilic head groups, as the strength of the interactions increased by the addition of small amounts of water. For water concentrations of more than 2%, the assembly was inhibited [32].

Vegetable waxes provide an ideal feature for organogels, which is more similar to the features provided by saturated and trans fats but show stability problems over prolonged periods. There is currently a clear lack of knowledge regarding the identification of ideal proportions of ceramides, fatty acids, cholesterols, and phytosterols, in terms of oil structuring, crystal morphology, and consequent suitability for applications in foods. Although carbohydrate-based gelatins were extensively studied, there are very few studies in the field of organogels [26].

4. Emerging crystallization technologies

The success of organogels in recent years in food applications continues to stimulate the interest of researchers in this area and provides a broader view [26].

A gel system is prepared by holding the solvent in the gelator matrix, stimulated by pH, heat, light, magnetic field, or ultrasound [47]. Supramolecular gels that have their structuring initiated by physical (including sonication) and mechanical (mechanical force) stimuli exhibit non-covalent interactions between the structurant molecules and show dynamic and reversible properties controlled by the stimuli, while structurants cause instant and *in situ* gelatinization in organic solvents or water with different modes and structuring results, but can lead to dynamic changes in the microscopic morphology, optical properties, etc. [45].

During the solvent-organogel transition process, changes in the conformation of the structure of the molecule can occur in thixotropic or ultrasonic organogels, but it is difficult to study these changes with the existing techniques and without crystallization data [45].

A novel low molecular weight, photoresistant organogel, based on an azobenzene derivative and which can achieve the reversible gel-solvent transition by the heat/cold process or visible/UV light irradiation, was designed and successfully synthesized. The structurant has a good gelation ability in organic solvents ranging from non-polar to polar solvents due to hydrogen bonds and π-π stacking [45].

The reversible control of the properties of the organogel through the application of heat and mechanical stimuli is often followed by transformations in the phase, morphological structure, and rheological and spectroscopic properties, which makes these organogels suitable to be used in the field of drug release and adaptive materials, among other applications [45].

5. Conclusion

This study carried out a comprehensive review on the possibilities of modification of the crystallization of lipid phases, to subsidize the knowledge of the crystallographic, microstructural, and kinetic phenomena involved in the processes of structuring oils and fats. Different emulsifiers used in low concentrations, besides the application of ultrasound technology, are proven effective in the processes of lipid modification and represent a highly feasible option, in economic and process terms, to modulate the crystallization properties of industrial oils and fats.

In addition, the use of organogels in food applications as substitutes for trans and saturated fatty acids is highly feasible since structured vegetable oils are generally used to replace fats. Compared with conventional technological processes for the production of technical fats for food applications, the production of organogels is technologically simple, economically accessible, and inexpensive. Several oils with majority composition in unsaturated fatty acids can be immobilized. Structuring agents are used in small proportions, being commercially available and safe for consumption.

Acknowledgements

This study was financed in part by the Coordination for the Improvement of Higher Education Personnel—Brazil (CAPES)—Finance Code 001.

Conflict of interest

The authors state that there is no conflict of interest.

Author details

Kamila Ferreira Chaves[1*], Thaís Jordânia Silva[1], Maria Aliciane Fontenele Domingues[2], Daniel Barrera-Arellano[1] and Ana Paula Badan Ribeiro[1]

1 School of Food Engineering, University of Campinas, Campinas, Brazil

2 Federal University of São Carlos, São Carlos, Brazil

*Address all correspondence to: chaves_kamila@yahoo.com.br

References

[1] Hashempour-Baltork F, Torbati M, Azadmard-Damirchi S, et al. Vegetable oil blending: A review of physicochemical, nutritional and health effects. Trends in Food Science and Technology. 2016;**57**:52-58. DOI: 10.1016/j.tifs.2016.09.007

[2] Rohm H, Schäper C, Zahn S. Interesterified fats in chocolate and bakery products: A concise review. In: Shahidi F, editor. LWT—Food Science and Technology. 2018;**87**:379-384. DOI: 10.1016/j.lwt.2017.08.076

[3] Ghotra BS, Dyal SD, Narine SS. Lipid shortenings: A review. Food Research International. 2002;**35**:1015-1048. DOI: 10.1016/S0963-9969(02)00163-1

[4] Sato K. Crystallization behaviour of fats and lipids—A review. Chemical Engineering Science. 2001;**56**:2255-2265. DOI: 10.1016/S0009-2509(00)00458-9

[5] Ribeiro APB, Masuchi MH, Miyasaki EK, et al. Crystallization modifiers in lipid systems. Journal of Food Science and Technology. 2015;**52**:3925-3946. DOI: 10.1007/s13197-014-1587-0

[6] Block JM, Barrera-Arellanos D. Temas Selectos en Aceites y Grasas. Química. Vol. 2. São Paulo, Brazil: Marcel Dekker, Inc, Blucher; 2013

[7] Roos YH. Phase and state transitions and related phenomena in foods. Food Materials Science and Engineering. 2012;**1**:1-15. DOI: 10.1002/9781118373903.ch5

[8] Hartel RW. Advances in food crystallization. Annual Review of Food Science and Technology. 2013;**4**:277-292

[9] Foubert I, Dewettinck K, Van de Walle D, et al. Physical properties: Structural and physical characteristics. In: The Lipid Handbook. Gunstone FD, Harwood JL, Dijkstra AJ, editors. 3rd ed. Boca Raton: CRC Press; 2007. pp. 47-81

[10] O'Brien RD. Oils and fats: Formulating and Processing for Applications. 3rd ed. New York: Taylor & Francis Group; 2008

[11] Patel AR, Dewettinck K. Current update on the influence of minor lipid components, shear and presence of interfaces on fat crystallization. Current Opinion in Food Science. 2015;**3**:65-70. DOI: 10.1016/j.cofs.2015.05.010

[12] Domingues MAF, Ribeiro APB, Kieckbusch TG, et al. Advances in lipids crystallization technology. Advanced Topics in Crystallization. 2015;**1**:105-132. DOI: 10.5772/59767

[13] Clercq N. Crystallization of fats and oils. In: KOI Cocoa Processing and Chocolate Production. Belgium: Ghent University; 2013

[14] Seddon KR, Zaworotko M. Crystal Engineering the Design and Application of Functional Solids. Holanda: Springer; 1999

[15] Hartman BP, Perdok WG. On the relations between structure and morphology of crystals. I. Acta Crystallographica. 1955;**8**:49-59. DOI: 10.1107/S0365110X55000121

[16] Fennema OR, William DP, Elmer HM. Low Temperature Preservation of Foods and Living Matter. Taylor & Francis; 1973

[17] Martini S, Awad T, Marangoni AG. Structure and properties of fat crystal networks. In: Gunstone F, editor. Modifying Lipids for Use in Food. Cambridge: Woodhead Publishing; 2006. pp. 142-169

[18] Garti N, Widlak NR. Cocoa Butter and Related Compounds. 1st ed. Urbana, Illinois: Elsevier; 2015

[19] Coultate TP. Food: The Chemistry of its Components. 6th ed. Cambridge: Royal Society of Chemistry; 2016

[20] Bayés-García L, Patel AR, Dewettinck K, et al. Lipid crystallization kinetics-roles of external factors influencing functionality of end products. Current Opinion in Food Science. 2015;**4**:32-38. DOI: 10.1016/j.cofs.2015.04.005

[21] Metin S, Hartel RW. Crystallization of fats and oils. In: Shahid F editor. Bailey's Industrial Oil and Fat Products. New York: Wiley Interscience; 2005. pp. 45-76

[22] Li Z, Cao J, Li H, et al. Self-assembled drug delivery system based on low-molecular-weight bis-amide organogelator: Synthesis, properties and in vivo evaluation. Drug Delivery. 2016;**23**:3168-3178. DOI: 10717544.2016.1157841

[23] Patrick M, Blindt R, Janssen J. The effect of ultrasonic intensity on the crystal structure of palm oil. Ultrasonics Sonochemistry. 2004;**11**:251-255. DOI: 10.1016/j.ultsonch.2004.01.017

[24] Castro MDL, Priego-Capote F. Ultrasound-assisted crystallization (sonocrystallization). Ultrasonics Sonochemistry. 2007;**14**:717-724. DOI: 10.1016/j.ultsonch.2006.12.004

[25] Chaves KF, Barrera-Arellano D, Ribeiro APB. Potential application of lipid organogels for food industry. Food Research International. 2018;**105**:863-872. DOI: 10.1016/j.foodres.2017.12.020

[26] Singh A, Auzanneau F, Rogers MA. Advances in edible oleogel technologies—A decade in review. Food Research International. 2017;**97**:307-317. DOI: 10.1016/j.foodres.2017.04.022

[27] Hughes NE, Marangoni AG, Wright AJ, et al. Potential food applications of edible oil organogels. Trends in Food Science & Technology. 2009;**20**:470-480. DOI: 10.1016/j.tifs.2009.06.002

[28] Moschakis T, Panagiotopoulou E, Katsanidis E. Sunflower oil organogels and organogel-in-water emulsions (part I): Microstructure and mechanical properties. LWT—Food Science and Technology. 2016;**73**:153-161. DOI: 10.1016/j.lwt.2016.03.004

[29] Garcia RKA, Gandra KM, Barrera-Arellano D. Development of a zero trans margarine from soybean-based interesterified fats formulated using artificial neural networks. Grasas y Aceites. 2013;**64**:521-530. DOI: 10.3989/gya.049113

[30] Marangoni AG, Garti N. Edible Oleogels: Structure and Health Implications. Amsterdam, Netherlands: Elsevier Inc; 2011. DOI: 10.1016/C2015-0-02413-3

[31] Sundram K, Karupaiah T, Hayes KC. Stearic acid-rich interesterified fat and trans-rich fat raise the LDL/HDL ratio and plasma glucose relative to palm olein in humans. Nutrition and Metabolism. 2007;**4**:3. DOI: 10.1186/1743-7075-4-3

[32] Nikiforidis CV, Gilbert EP, Scholten E. Organogel formation via supramolecular assembly of oleic acid and sodium oleate. RSC Advances. 2015;**5**:47466-47475. DOI: 10.1039/c5ra05336f

[33] Uzan S, Barış D, Çolak M, et al. Organogels as novel carriers for dermal and topical drug delivery vehicles. Tetrahedron. 2016;**72**:7517-7525. DOI: 10.1016/J.TET.2016.10.009

[34] Mandal D, Mandal SK, Ghosh M, et al. Phenylboronic acid appended pyrene-based low-molecular-weight injectable hydrogel: Glucose-stimulated insulin release. Chemistry—A European Journal. 2015;**21**:12042-12052. DOI: 10.1002/chem.201501170

[35] Zhao Y, Zhou L, Liu J, et al. Preparation and investigation of a novel levobupivacaine in situ implant gel for prolonged local anesthetics.

Artificial Cells, Blood Substitutes, and Biotechnology. 2017;**45**:404-408. DOI: 10.3109/21691401.2016.1160406

[36] Hu B, Wang W, Wang Y, et al. Degradation of glutamate-based organogels for biodegradable implants: In vitro study and in vivo observation. Materials Science and Engineering: C. 2018;**82**:80-90. DOI: 10.1208/s12249-018-1074-4

[37] Long D, Gong T, Zhang Z, et al. Preparation and evaluation of a phospholipid-based injectable gel for the long term delivery of leuprolide acetaterrh. Acta Pharmaceutica Sinica B. 2016;**6**:329-335. DOI: 10.1016/j.apsb.2016.05.004

[38] Martin B, Brouillet F, Franceschi S, et al. Evaluation of organogel nanoparticles as drug delivery system for lipophilic compounds. AAPS PharmSciTech. 2017;**18**:1261-1269. DOI: 10.1208/s12249-016-0587-y

[39] Pernetti M, Vanmalssen K, Kalnin D, et al. Structuring edible oil with lecithin and sorbitan tri-stearate. Food Hydrocolloids. 2007;**21**:855-861. DOI: 10.1016/j.foodhyd.2006.10.023

[40] Lupi FR, Greco V, Baldino N, et al. The effects of intermolecular interactions on the physical properties of organogels in edible oils. Journal of Colloid and Interface Science. 2016;**483**:154-164. DOI: 10.1016/j.jcis.2016.08.009

[41] Huang CB, Chen LJ, Huang J, et al. A novel pyrene-containing fluorescent organogel derived from a quinoline-based fluorescent porbe: Synthesis, sensing properties, and its aggregation behavior. RSC Advances. 2014;**4**:19538-19549. DOI: 10.1039/c4ra02373k

[42] Pirner D, Dulle MEJ, Mauer M, et al. Reinforcement of nanostructured organogels by hydrogen bonds. RSC Advances. 2016;**6**:42730-42738. DOI: 10.1039/c6ra03567a

[43] Simsolo EE, Eroglu I, Tanriverdi ST, et al. Formulation and evaluation of organogels containing hyaluronan microparticles for topical delivery of caffeine. AAPS PharmSciTech. 2018;**19**:1367-1376. DOI: 10.1208/s12249-018-0955-x

[44] Steed JW. Supramolecular gel chemistry: Developments over the last decade. Chemical Communications. 2011;**47**:1379-1383. DOI: 10.1039/c0cc03293j

[45] Yu X, Chen L, Zhang M, et al. Low-molecular-mass gels responding to ultrasound and mechanical stress: Towards self-healing materials. Chemical Society Reviews. 2014;**43**:5346-5371. DOI: 10.1039/c4cs00066h

[46] Co ED, Marangoni AG. Organogels: An alternative edible oil-structuring method. Journal of the American Oil Chemists' Society. 2012;**89**:749-780. DOI: 10.1007/s11746-012-2049-3

[47] Hu B, Sun W, Yang B, et al. Application of solvent parameters for predicting organogel formation. AAPS PharmSciTech. 2018;**19**:2288-2300. DOI: 10.1208/s12249-018-1074-4

[48] Hoffmann H. Fascinating phenomena in surfactant chemistry. Advances in Colloid and Interface Science. 1990;**32**:123-150. DOI: 10.1016/0001-8686(90)80015-R

[49] Gennes PG. Scaling Concepts in Polymer Physics. London: Cornell University Press; 1979

[50] Gelbart WM, Ben-Shaul A. The 'new' science of 'complex fluids. The Journal of Physical Chemistry. 1996;**100**:13169-13189. DOI: 10.1021/jp9606570

[51] Sonin AS, Churochkina NA. Liquid crystals stabilized by physical networks.

Polymer Science, Series A. 2013;**55**:353-384. DOI: 10.1134/S0965545X13060060

[52] Hirst AR, Smith DK, Harrington JP. Unique nanoscale morphologies underpinning organic gel-phase materials. Chemistry—A European Journal. 2005;**11**:6552-6559. DOI: 10.1002/chem.200500501

[53] Wu S, Gao J, Emge TJ, et al. Solvent-induced polymorphic nanoscale transitions for 12-hydroxyoctadecanoic acid molecular gels. Crystal Growth and Design. 2013;**13**:1360-1366. DOI: 10.1021/cg400124e

[54] Imai M, Hashizaki K, Taguchi H, et al. A new reverse worm-like micellar system from a lecithin, multivalent carboxylic acid and oil mixture. Journal of Colloid and Interface Science. 2013;**403**:77-83. DOI: 10.1016/j.jcis.2013.04.033

[55] Nikiforidis CV, Scholten E. Self-assemblies of lecithin and α-tocopherol as gelators of lipid material. RSC Advances. 2014;**4**:2466-2473. DOI: 10.1039/c3ra46584e

[56] Lan Y, Corradini MG, Weiss RG, et al. To gel or not to gel: Correlating molecular gelation with solvent parameters. Chemical Society Reviews. 2015;**44**:6035-6058. DOI: 10.1039/C5CS00136F

[57] Haldar S, Karmakar K. A systematic understanding of gelation self-assembly: Solvophobically assisted supramolecular gelation via conformational reorientation across amide functionality on a hydrophobically modulated dipeptide based ambidextrous gelator, N-n-acyl-(l)Val-X(OBn). RSC Advances. 2015;**5**:66339-66354. DOI: 10.1039/c5ra10209j

[58] Zhao C, Wang H, Bai B, et al. Organogels from unsymmetrical π-conjugated 1,3,4-oxadiazole derivatives. New Journal of Chemistry. 2013;**37**:1454-1460. DOI: 10.1039/c3nj40648b

[59] Edwards W, Smith DK. Dynamic evolving two-component supramolecular gels-hierarchical control over component selection in complex mixtures. Journal of the American Chemical Society. 2013;**135**:5911-5920. DOI: 10.1021/ja4017107

[60] Bustamante P, Navarro-Lupión J, Peña MA, et al. Hildebrand solubility parameter to predict drug release from hydroxypropyl methylcellulose gels. International Journal of Pharmaceutics. 2011;**414**:125-130. DOI: 10.1016/j.ijpharm.2011.05.011

[61] Bonnet J, Suissa G, Raynal M, et al. Organogel formation rationalized by Hansen solubility parameters: Dos and don'ts. Soft Matter. 2014;**10**:3154-3160. DOI: 10.1039/c4sm00244j

[62] Bonnet J, Suissa G, Raynal M, et al. Organogel formation rationalized by Hansen solubility parameters: Influence of gelator structure. Soft Matter. 2015;**11**:2308-2312. DOI: 10.1039/c5sm00017c

[63] Trickett K, Eastoe J. Surfactant-based gels. Advances in Colloid and Interface Science. 2008;**144**:66-74. DOI: 10.1016/j.cis.2008.08.009

[64] Israelachvili J. Intermolecular and Surface Forces. 3rd ed. USA: Elsevier; 2011

[65] Suzuki M, Hanabusa K. Polymer organogelators that make supramolecular organogels through physical cross-linking and self-assembly. Chemical Society Reviews. 2010;**39**:455-463. DOI: 10.1039/b910604a

[66] Dastidar P. Supramolecular gelling agents: Can they be designed? Chemical Society Reviews. 2008;**37**:2699-2715. DOI: 10.1039/b807346e

[67] Zhang Z, Zhang S, Zhang J, et al. Solvent-dependent self-assembly and morphological transition of

low-molecular-weight azobenzene organogel. Tetrahedron. 2017;**73**: 4891-4895. DOI: 10.1016/j.tet.2017.05.027

[68] Sahoo S, Kumar N, Bhattacharya C, et al. Organogels: Properties and applications in drug delivery. Designed Monomers and Polymers. 2011;**14**: 95-108. DOI: 10.1163/138577211X555721

[69] Daniel J, Rajasekharan R. Organogelation of plant oils and hydrocarbons by long-chain saturated FA, fatty alcohols, wax esters, and dicarboxylic acids. Journal of the American Oil Chemists' Society. 2003;**80**:417-421. DOI: 10.1007/s11746-003-0714-0